D0333710

# THE OLIVETTI CHRONICLES

www.rbooks.co.uk

*Also by John Peel*

Margrave of the Marshes

# THE OLIVETTI CHRONICLES

## Three Decades of Life and Music

John Peel

BANTAM PRESS

LONDON • TORONTO • SYDNEY • AUCKLAND • JOHANNESBURG

TRANSWORLD PUBLISHERS
61–63 Uxbridge Road, London W5 5SA
A Random House Group Company
www.rbooks.co.uk

First published in Great Britain
in 2008 by Bantam Press
an imprint of Transworld Publishers

Copyright © Sheila Ravenscroft 2008

The right of Sheila Ravenscroft to be identified as the
author of this work has been asserted under the Copyright,
Designs and Patents Act 1988.

A CIP catalogue record for this book
is available from the British Library.

ISBN 9780593060612 (cased)
9780593062142 (tpb)

This book is sold subject to the condition that it shall not,
by way of trade or otherwise, be lent, resold, hired out,
or otherwise circulated without the publisher's prior
consent in any form of binding or cover other than that
in which it is published and without a similar condition,
including this condition, being imposed on the
subsequent purchaser.

Addresses for Random House Group Ltd companies outside the UK
can be found at: www.randomhouse.co.uk
The Random House Group Ltd Reg. No. 954009

The Random House Group Limited supports The Forest Stewardship
Council (FSC), the leading international forest-certification organisation.
All our titles that are printed on Greenpeace-approved FSC-certified
paper carry the FSC logo. Our paper procurement policy can be found at
www.rbooks.co.uk/environment

Typeset in 11/15.5pt Berling by
Falcon Oast Graphic Art Ltd
Printed and bound in Great Britain by
Clays Ltd, Bungay, Suffolk

2 4 6 8 10 9 7 5 3 1

Every effort has been made to obtain the necessary permissions with
reference to copyright material, both illustrative and quoted.
We apologise for any omissions in this respect and will be pleased
to make the appropriate acknowledgements in any future edition.

# Contents

| | |
|---|---|
| Acknowledgements | ix |
| Introduction | xi |
| | |
| A Snappy Dresser | 1 |
| Alien Empire | 4 |
| Archers | 7 |
| Aural Vandalism | 9 |
| Babies | 12 |
| Bembeya Jazz National | 15 |
| Tony Bennett | 17 |
| Berlin Punk | 19 |
| Bhangra | 22 |
| The Bhundu Boys | 25 |
| Bores | 27 |
| Bottom Lines | 30 |
| Butthole Surfers | 32 |
| California | 34 |
| Captain Beefheart | 36 |
| Captain Beefheart 2 | 39 |
| Chicago House | 42 |
| Children's TV | 44 |
| Kenny Dalglish | 47 |
| Devil's Music | 49 |
| Doomsday | 52 |
| Driving to Sonar | 55 |
| Eastern Bloc | 57 |
| Duane Eddy | 61 |
| Europe | 64 |
| Eurovision | 66 |

Eurovision 2 69
Everything's Up to Date in Kansas City 73
Extreme Noise Terror 75
Fab Pic Contest! 77
Fab Pic Contest 2 80
Fab Pic Contest 3 84
Faces 88
The Fall 90
Bryan Ferry 92
Football 94
Foxes 97
Glastonbury 100
God Save the Queen 104
Half Man Half Biscuit 108
Happy Mondays 110
Hello Tailor! 112
Adrian Henri 114
Hippies 117
Ipswich 119
Michael Jackson 122
Billy Joel 125
Kerguelen 127
King Arthur 130
Knebworth 133
Kosmische Musik 136
Frankie Laine 139
Liverpool 141
Local Radio 143
Lovelace 146
Madonna 151
Madonna 2 154
Medieval Medicine 156
Kylie Minogue 158
Misty in Roots 160

Montel 163
The Mourning of the Golden Flask 166
Napalm Death 169
New Age Music 171
New Year's Eve 173
1975 175
1977 180
Nooderslag 186
Oddballs 189
Old Bill 191
Roy Orbison 195
Osmonds 197
Osmonds 2 201
Osmonds 3 204
The Party's Over 208
Peel and the Mighty Gorgon 211
Peel at the 'Quiet' Albert Hall 214
Peel's Theory 218
Pen Pal 220
Phoenix Festival 223
Pink Pop 226
Please Listen to My Programme 230
Prince 233
Protest Songs 235
Public Enemy 237
Radio 1 239
Reading Festival 242
Reading Festival 2 245
Record Shops 251
Cliff Richard 255
Road Rage 257
Roadshows 260
Roadshows 2 263
Rock's in Trouble 266

| | |
|---|---|
| Rockpool | 268 |
| Shaving | 270 |
| Sick in Trains | 273 |
| The Smiths | 276 |
| Sock Syndrome | 278 |
| Sound City | 281 |
| Viv Stanshall | 284 |
| Viv Stanshall 2 | 286 |
| Sub Pop | 290 |
| Summer of Love | 292 |
| Today Programme | 295 |
| Tommy | 297 |
| Too Decrepit to Walk | 299 |
| Top of the Pops | 301 |
| TT Races | 304 |
| Tubular Bells | 311 |
| Turntable Mistress | 314 |
| UK Fresh 86 | 316 |
| USSR | 318 |
| Virginity | 320 |
| Voice-overs | 324 |
| Loudon Wainwright | 327 |
| John Walters | 331 |
| Wham! | 333 |
| Whatever You Want | 336 |
| WI | 339 |
| Wine-tasting with Walters | 341 |
| World Service | 345 |
| Robert Wyatt | 347 |
| Lena Zavaroni | 350 |
| | |
| Picture Credits | 353 |
| Index | 355 |

# Acknowledgements

We would like to thank Paul Bignell, Jennifer Forde, Matt Gocher, Annie Hatt, Andy Kershaw, Phil Knappett, Cat Ledger, Zahra Ravenscroft, Sue Robinson, Clive and Shurley Selwood, Roger Tooth, Mariam Yamin; Doug, Rebecca, Alison and all the staff at Transworld; and all those we have almost certainly forgotten to mention.

Thanks also to *BIKE* magazine, *Disc, Guardian, Independent on Sunday, The Listener, Observer, Punch, Radio Times* and *Sounds* for originally publishing John's material.

# Introduction

Dad, a sentimental man, insisted on using his Olivetti typewriter to write the majority of the articles contained in this book, despite the fact that it was so ancient that he could rarely find ink for it. As the supply of specially ordered compatible ink ribbons ran dry, he discovered a way to messily rethread the old ones and put them in backwards, reusing them until what he had written was barely visible.

Looking over these decades' worth of newspaper and magazine columns whilst fact-finding for *Margrave of the Marshes*, we came to realise how well observed and entertaining most of them were. This is a book we put together as much as anything else because we each wanted to own a copy. Happily, we don't seem to be alone in this. Many people have suggested to us since Dad's death in 2004 that we compile some of the best of his writing – so here it is.

We've tried to put together a diverse selection of columns, starting with some undated *Disc* pieces from 1971, we think. There were earlier works available to us but Dad wouldn't have thanked us very much for reproducing any of his articles from *Gandalf's Garden* or *International Times*:

> Walk far from the towns. Touch the bark of a thousand trees, shoeless. There are a million flowers that have grown just for

you. Please go and find them and let them know your peace . . . Go to the children's playground in Kensington Gardens and stare at the elves on the trees there.

The editions of *Disc, Sounds, The Listener, Punch, Observer, Guardian, BIKE* magazine, *Independent* and *Radio Times* that we have plundered include pieces covering a great many subjects – record shopping, Eurovision, Dad getting his corns trimmed, football, a lot of gig reviews and talk about how everyone should be listening to his programme, and how ill he's feeling.

You'll find that the subjects and chronology jump about quite a bit. We opted for alphabetical order by title so that we could engineer a bit of balance to the book. We hope this doesn't upset the flow, but that it creates a bit of variety and highlights some change over the years, sometimes a lack of it. Many of the older articles are alarmingly relevant now ('Rock's in Trouble', for example). There are a few instances where what Dad has written doesn't seem to make much sense, and we're not quite sure what he meant to say. These have been left as little puzzles.

Hopefully Dad would have approved of our having gathered all of this up and brought it to your attention. He often expressed mild embarrassment at some of the things he used to write, but this seems to us to be utterly without foundation. We think of him every day, and are very grateful that we've been able to spend the best part of a year hearing his voice in these articles, and we very much hope you'll enjoy them as much as we have.

William Ravenscroft

# A Snappy Dresser

*Radio Times*, 13–19 June 1998

---

I LEFT London's Grosvenor House Hotel feeling strangely buoyant. Outside on Park Lane the air smelt sweet despite the late-night traffic. On my arm was a beautiful woman I had met a scant twenty-nine years earlier. But let me start at the beginning . . . Born in the woodcutter's cottage at the age of three, John Peel was . . . Well, perhaps that's going back too far. Let's settle instead for . . . The invitation had arrived some seven or eight weeks before the events that were to culminate in the Night of the Big Trouser. 'Dear (YOUR NAME HERE),' it began. 'I hope you can join us,' it continued, 'for the 1998 Periodical Publishers Association awards dinner. It's a black tie event.' Reading on, I learnt that I had been nominated for Columnist of the Year (Consumer Magazines) for this very column.

'Darling,' I shouted up the stairs, 'I've been nominated for Columnist of the Year Open Brackets Consumer Magazines Close Brackets.'

'Can't hear you,' came the answering cry.

I'm not a snappy dresser. I can see that now. I do own a suit but when I can be persuaded into it, I look like a rather timid bouncer in one of the less successful eastern European countries – Albania, perhaps. I don't own evening dress and, what with one thing and another, left it too late to hire such an outfit. At the last minute, after a bizarre measuring ritual in our kitchen that may well become an annual event in its own right, my personal dimensions were phoned to a secret location in London and a suit duly dispatched to our luxury hotel.

Having spent the day pre-recording programmes and practising my gracious loser's face in the mirror, my first meeting with the suit came less than an hour before the Periodical Publishers fell to their banqueting and, as I unfolded the trousers on to the hotel bed, it became clear that here was a garment intended for community use rather than for the use of one selfish individual.

Frankly, left to my own devices I would have spent the evening watching television and concocting an elaborate excuse, possibly involving alien abduction, for my absence. However, Sheila, with many a honeyed, 'No, you look fine as long as you keep the jacket buttoned,' and 'Honestly, they're not laughing at you,' wheedled and bullied me out of our room, out of the hotel and into the street.

An hour later, as I toyed listlessly with my food and the trousers billowed in the blasts of air from the air-conditioning system, I imagined the shouts of laughter that would greet my walk to the stage if I were to be named Columnist of the Year (Consumer Magazines). So, when top TV funnyman Ian Hislop had read the names of the nominees, the dread words '. . . and the winner is . . .' had been spoken, and a name other than mine (James May of *Car* magazine) had been announced, a curious sense of wellbeing spread over me.

As the evening wore on, as evenings will, the uproar from the BBC Magazines tables grew, culminating in a great shout of triumph when *RT*'s stablemate *Top of the Pops* was named Magazine of the Year. Understandably, the Top of the Popsters, average age about fifteen, went right off the rails and stayed up long after they should all have been in bed. They were still celebrating when Sheila and I stepped into the street, both of us just faces in the crowd, neither of us columnists of this or any other year – although Sheila did once win a meal for two at an Italian restaurant in Ipswich and I won a tin of toffees at a grocery shop in Neston when I was about fifteen myself.

This was to have been the year when I finally went to a World Cup match – any World Cup match – but the bizarre ticketing arrangements have put paid to that. With an England team of bewitched veterans and a Scotland without Liverpool players, domestic support may well switch to Jamaica.

A democratic decision will be reached after we've all watched *The Reggae Boyz* documentary on Saturday on Channel 4.

# Alien Empire

*Radio Times*, 17–23 February 1996

PITY THE POOR mayfly. It starts life in riverbottom mud, reaches the surface with no means of eating, has one day of life, then dies. This is a process that has gone on, without so much as a murmur of complaint, for over 100 million years. I know this from watching *Alien Empire: Replicators*, narrated by John Shrapnel.

I expect that you rather like, as I do, those Strange Facts that crop up in programmes such as this. You know the sort of thing. If all the people who have ever written to me to ask for the title of that long reggae record I played a few years ago – it might have been in August – or March – were gathered together, there would be enough of them to populate at least three of the sun-dappled villages that dot the countryside around Aix-en-Provence.

There's quite a bit of this sort of thing in *Alien Empire*. For example, if two flies – any two, it doesn't matter which, provided, of course, that one is male, the other female – started mating and were allowed to go on doing this unhindered, in other words, without any of their great-great-great-great-great-great-great-great-great-great-great-great-great-great-great-great-great-great-great-great-great grandchildren getting tangled up in one of those blue light things, there would be, at the end of the year, enough flies to make a ball as big as the earth. Something like that anyway. Let's just hope it doesn't happen, eh?

We humans tend to be pretty smug about some of the tricks we can do. We do have the means of eating, for starters. We can stick cotton buds in our ears, against medical advice, and get out satisfying quantities of ear wax. Some of us can watch Jim Davidson on television without contemplating starting a new life among the mountain peoples of Peru. But can we release scent from a gland in our abdomens, a scent so magnificent that a potential partner can smell it over a mile away and come scurrying to our sides with a soft grin on his face? I think not. There is some sort of moth in the USA that thinks nothing of doing this.

Can you, as the polka-dot moth can, hear in ultrasound? Can you say with any certainty what ultrasound is? Wasn't it the system that went with Todd-AO?

If you are less than impressed with these moths and their fancy ways, consider the leaf-hopper. I rather took to the leaf-hopper, I must admit. It is short, stubby and the colour of our bathroom carpet. It also – and I recommend that you clutch at a piece of furniture before reading further – hears with its feet. This is a good trick in any school.

I won't spoil your enjoyment of *Alien Empire* by spilling more of the programme's beans. Well, some backbenchers might like to know that single mothers in the wasp community – that's wasp, not WASP – nourish their young at someone else's expense rather than employ a child minder and go out and find work, of which there is plenty if you could just be bothered to look, oh yes there is . . .

Oh, and jewel beetles in Australia fall in love with beer bottles – and John Shrapnel gets to read such lines as, 'Garden plants (are) alive with secret music.'

For a fortnight, I have been sending messages to the *RT* offices in London (from a gland in my abdomen, since you ask) in an attempt to find out why I was not told in advance of the recent

*Under the Sun: Painted Babies.* You may recall that five-year-old Brooke Breedwell was winner of the Universal Southern Charm Beauty Pageant in Atlanta, Georgia, and that there was a cut out 'n' keep picture of her on page 23 of the *Radio Times* of 27 January.

I don't want to be unkind to a five-year-old, but Brooke had a look so steely that a single glance would probably be enough to stop the mighty, meandering Mississippi as it winds its centuries-old way to the Gulf of Mexico and the open sea beyond. One can imagine that Brooke, groomed and groomed and groomed again for pageant after pageant, might well butcher her parents when she is old enough to carry arms – say, when she is seven or eight.

Then local legal traditions will take over. An Afro-Caribbean American, preferably a handyman who was drinking with eleven of his pals in a neighbouring county at the time of the killing, will be arrested, tried, convicted and kept on death row for seventeen years before being electrocuted. Maybe being a mayfly isn't so bad after all.

# Archers

*Guardian*, 28 October 1994

---

TURN, IF YOU WILL, to page 231 of *The Book of the Archers*, put together by Patricia Greene (Jill Archer), Charles Collingwood (Brian Aldridge) and Hedli Niklaus (Kathy Perks), and you will find a twenty-four-line entry under Peel, John. Yes, it's me all right.

There was a time when you could have delighted guests at smart dinner parties by asking: 'What do Princess Margaret (8 lines), the Duke of Westminster (15 lines), the former Lucinda Prior-Palmer (no lines at all) and Radio 1's Nabob of Noise, John Peel, have in common?' The answer would have been that we were the only souls on Earth to have played ourselves in *The Archers*. (I don't know what happened to Lucinda. Perhaps she wasn't in it after all.)

In more recent times, alas, the tone has been lowered more than a little by the addition of such names as Britt Ekland (16 lines), Anneka Rice (6 lines) and Terry Wogan (9 lines).

We of the original inner circle seldom have time to get together these days, but I did once find myself at a knees-up to celebrate some sort of *Desert Island Discs* anniversary, discussing our roles in *The Archers* with Princess Margaret. I don't believe I betray any especial confidences if I tell you that the Princess graciously admitted to having enjoyed being featured in the long-running everyday story of country folk. Since I have initiated a low-key campaign to give the Stuarts another go at being the Royal Family, thereby allowing the present lot a century or two in the reserves in which to rediscover their form, I doubt we will meet again to talk of Ambridge.

Actually, I have featured in *The Archers* on two distinctly separate occasions. On the first, I was heard on the radio 'on the radio', as it were. Eddie Grundy and the hapless Clarrie were listening to Radio 1 in the kitchen at Grange Farm when I played one of Eddie's records. On the second, my work was spread over several episodes and I even got to sing 'Yellow Submarine' in a van with Eddie and that nice Mrs Antrobus.

This time, I discovered that there is rather more to this acting game than I had imagined and my conviction that I had delivered my lines with character and a certain brio was shown to be mere self-delusion when the programmes were heard. Alexandra, then aged about twelve, scampered into the kitchen as the rest of us clustered about the radiogram, listened for a second, then said, with the careless cruelty of the young: 'You can tell you're reading it, Dad.' This, coming as it did from someone who watches *Neighbours* and *Home and Away* without a qualm, was a criticism too far and with it I turned my back on the theatre. There are those fanatics who have listened to *The Archers* since the programme was first heard in 1951 and could tell you at once who, for example, Mrs Bradshaw (page 81, 3 lines) was, but I am not one of those. Listening from time to time is enough to keep abreast of major developments, although I am sufficiently engaged to wish Kate Aldridge would phone her mother and put an end to all our suffering.

Come on, Kate, you know it makes sense. I'll get Pete Tong to play you a record if you do.

# Aural Vandalism

## Sonic Youth and The Jesus and Mary Chain

*Observer*, 11 May 1986

---

THE CULTURE SHOCK involved on Thursday in making the transition from the amiable tomfoolery of *Top of the Pops* to the Hammersmith Palais for Sonic Youth and The Jesus and Mary Chain was as considerable as any I have experienced.

Simon Frith, in a memorable phrase, said of Sonic Youth that these are 'clever people playing stupid music'. I cannot confirm or deny their cleverness but I do not believe that their music is stupid. At a time when vandalism in the streets, vexing and dangerous as it can be, is seeming more and more to be a reasonable response to the circumstances in which far too many people live, Sonic Youth's music is nothing more than aural vandalism, designed to disturb and confront.

A confrontational approach is applauded in the cinema, the theatre and the gallery, why not on the dance floor? I was even slightly disappointed at not being as affronted and affrighted by the band's performance as I had expected. It has been said that Sonic Youth are keen to recapture the energies that transformed American rock in the late 1960s and, on stage in Hammersmith, hunched and anonymous, all feedback and purple lighting, they certainly recaptured much of the spirit I recollect from Sunset Strip in 1967.

As with the very best bands, Sonic Youth's lyrics convey quick impressions and half images. I caught the line 'the size of the towtruck', almost certainly misheard, and have been repeating it like a mantra ever since.

The American quartet left the stage with guitars still shrieking – a very sixties touch – then made the mistake of playing an encore. I am not, I confess, an encore man. I doubt that Leonardo, having completed work on *The Last Supper*, presumably to some local acclaim, would have interpreted this acclaim as a signal to paint in a few more figures, and find the utterly predictable making and taking of encores somehow rather debasing.

The Jesus and Mary Chain, for all their apparent naivety, know about encores. Their brief history has been composed largely of rumour, speculation and conflict. We read that they are older than they claim, that their live performances often end in disarray, and that they have played on occasion for less than twenty minutes. If twenty minutes is the time it takes The Jesus and Mary Chain to achieve whatever effect it is they wish, it would be fatuous to play longer. Music should not be, although I fear that it often is, something that you buy by the yard. In Hammersmith they played for forty minutes.

John Foster-Moore, the new drummer, plays standing up, another echo of the sixties, and in a room swooning with the sweet smell of radical hair, The Jesus and Mary Chain's often poignant songs were as affecting as they are on record, the ends of half-whispered words whipped away in gales of feedback. The newer songs, the titles of which were neither given nor needed, are less wistful, somewhat crueller, reflecting perhaps the lives this much maligned band now live.

It is so easy to romance about new bands, to credit the dull, unhappy cities and towns that spawn them (The Jesus and Mary Chain are from East Kilbride) with shaping the music, but I believe that such bands are playing real folk music, in the sense that their work fairly represents their experience and that that experience is universal.

As I left the Palais I overheard one twerp braying to his companions. 'What a waste of money,' he cried, and for a moment I wished I were a fighting man.

# Babies

*Sounds*, 14 August 1976

---

YES, I REMEMBER how sensitive you are, and normally, as I'm sure you're aware, I wouldn't think of mentioning it. But – don't turn around right now – he's still lying there on the tiles and staring up at me.

From the expression on that part of his face which is visible beneath layers of perfectly nauseating baby breakfasts, he can hardly wait to jump to his feet and visit violence upon his cringing father. Fortunately, according to the best works on the subject, it will be some months before he can begin to think seriously about translating such thoughts into deeds.

I have, I'm afraid, been left alone in the house with William (A Baby). At the moment he is laughing hysterically over, as far as one can tell, nothing at all. From time to time he unleashes a shriek of pleasure so pure and high-pitched that all the creatures in the adjoining hedgerows pause and look at one another, lost in admiration at such a sound. At night the impressionable mind – which is what your Uncle John comes equipped with – can easily persuade itself that this single frightful call is summoning the shades of the long dead; and sitting bolt upright in bed, white-faced and trembling, in the wee small hours of the morning, has become a regular part of our lives.

William still hasn't mastered the art of forward motion, but he can now fix his beady eye on a target and move backwards towards it. At this instant he has set his sights on Mile (A Dog), on whom he seems anxious to demonstrate some infant martial art. Mile (A Dog) looks worried and leaves for calmer fields.

Five minutes later. I have decided to try out the new Grateful Dead LP, *Steal Your Face*, on him. He doesn't seem to care for it a lot and one can see what he means. The vocals are awful. Hold on, let's see how he reacts to the new Joan Armatrading album – I expect you heard the world exclusive single track that Fluff, a longtime Armatrading fan, played the Saturday before last. You didn't? Oh, well, never mind.

I know Loudon Wainwright has already mentioned it in song, but babies are extraordinarily oral little devils, are they not? William has already sucked most of the probably harmful dye out of the Liverpool FC shield stuck inside his pram, and seems to have set himself the larger task of gumming the actual body of the pram to smithereens. But enough of babies . . . what I had intended to say was this . . .

POINT ONE. You probably made less sense than usual of last week's column. This, through no fault of the delightful creature who took it down over the 'phone, became considerably altered between leaving my head and reaching yours. In fact, there were passages that I couldn't begin to fathom either. The point I was keen to convey was that I'm about to go a-holidaying-o in Devon and that while I'm away we will be running some special programmes on Wunnerful Radio 1, your friendly motor-sports station, every night from eleven to midnight, VHF, stereo, 247 metres and so on.

You will have already missed the first two or three, which will have dealt with the music of The Who, Family and Roxy Music, and of individual members of those bands, but I will not listen to excuses for missing the remaining programmes which, this week, deal pretty firmly with Fairport Convention and the Stones and, next, with Cream, the Softs, the Faces, the Yardbirds, and the Fab Four. Questions will be asked after the series and if you score less than 85 per cent you'll be kept in after school. You, Deidre, will be kept in anyway, and you can help me tidy up the cricket pavilion.

(For those of you who care, William (A Baby) has fallen asleep, albeit in a position I couldn't achieve in fifty years of study on the banks of the Ganges. The Grateful Dead have their uses, after all.)

Actually, it is sad to see bands, such as the Dead, who have contributed so much towards the enrichment of our musical lives, making such low-level recordings.

I had a letter recently from a disgruntled viewer who asked what had happened to my radio programmes. 'Why', he sought to discover, 'do you no longer play the great records you used to play – the Dead, Airplane, Incredible String Band and such?'

What could I say, my little ones? Only quote that old adage about the thingummy of the music changing. You know the one, I'm sure. Now the energy is more widely spread, of course, and there is dandy music to be heard in all sorts of places.

In fact, only last night I ambled along, together with Gabs, ginger-haired sister-in-law by appointment to Peel Acres, to visit with the Be-Bop Deluxes as they slaved over their forthcoming LP. Unfortunately I made the mistake, as I often do, of talking far too much, and only heard a couple of tracks from the album. Both sounded, you will rejoice to learn, well up to their uncommonly high standard.

The only half-way sensible thing I said as I prattled on was that you could take five seconds – any five seconds – from any of their records and it would be immediately identifiable as being by Be-Bop Deluxe. That, I concluded triumphantly, is one of the marks of a great band. And so it is.

Next week's bulletin will come by Her Majesty's mails from Padstow – we'll have a whipround for the postage – and will almost certainly have at least as little to do with music as this lot.

# Bembeya Jazz National

*Observer*, 26 July 1987

---

THERE ARE TIMES – are there not – when the animal passions are so engaged that you cannot understand why the breath heaving from your flaring nostrils fails to set your shirt alight. We are talking here about sex.

I am not myself, I reluctantly admit, much given to or sought after for Hunnish practices – an estate agent attempting to interest clients in my oh-so-subtle charms might write something like 'Rural property in need of modernisation. Delightful views. Some damp.' But the music of Bembeya Jazz National, visiting the Africa Centre in Covent Garden from Guinea-Conakry last week, struck me as being impudently erotic.

You will, without doubt, recollect previous jottings on the subject of African music in these pages, how hardly a year passes in which we are not assured by some authority or other that the aforementioned African music is poised – I think that is the word they use – to make significant inroads into mainstream pop. It never happens, of course, and it never will, but during 1986–87 Zimbabwe's Bhundu Boys, followed by Jonah Moyo and the Devera Ngwena Jazz Band and Real Sound, have, in conjunction with a range of home-grown bands, made the less entrenched British audiences at least aware of the sound of African music by the simple means of playing it often and playing it well.

'African music' is clearly as meaningful or as meaningless as 'European music' or 'Asian music', but there are common characteristics, principally and obviously a considerable but flexible rhythmic drive and an abundance of free-flowing electric guitar playing. At the Africa Centre, Bembeya demonstrated both of these characteristics marvellously well.

Bembeya Jazz National is, even by local standards, a well-established band. Sekou Diabate, known to admirers as Diamond Fingers, was adjudged Africa's best guitarist in 1977 and the équipe itself, if my translation of the French so recklessly employed on the sleeve of a recent LP is correct, has been at it since 1961.

Fielding two trumpets, a tenor sax, guitar, bass, rhythm guitar, drums and percussion, along with three singers decked out in sporty red-and-white matching outfits, Bembeya played to a disappointingly small crowd in Covent Garden. However, those in attendance were not down-hearted. The Guineans play music which has evolved, I am assured, from West African Mandinka rhythms and is called mbalax, a fact which I derived some obscure pleasure communicating to my radio audience, and as they do it and melody upon melody flows from the guitar of Sekou Diabate, the overall effect, as I shamefacedly suggested above, hits below the belt.

In Bembeya's music, as in all the best music whatever its source, there is a considerable sense of space. Rhythms are implied rather than relentlessly stated and somehow the listener's heartbeat seems to fill these gaps. Yet again the pages of my reporter's notebook remained unsullied as I closed my eyes and cursed my inability to dance. Is hypnotherapy a possible solution here?

# Tony Bennett

*Radio Times*, 12–18 May 2001

---

'WHAT A NICE man he seems,' sighed Sheila as Tony Bennett brought BBC1's recent Judy Garland documentary to a simple, sad conclusion. I agreed with her. Not that we've ever met Tony Bennett, you understand, although I was on stage when he performed at a recent Glastonbury festival.

The singer and a handful of gifted musicians who had, I assumed, worked with him for aeons and had discovered long since that you don't have to play flat out all the time for maximum effect, were in the process of giving us their full club act. Idly – how else? – I wondered what the urbane, immaculate American had thought when he stepped on stage and saw an audience of gurning mudlarks stretching to the gun-metal grey horizon. Whatever his initial response may have been, Tony seemed moved by the enthusiastic reaction of the Glastonbury fun-seekers, some of whom were attempting to foxtrot in the mud. So much so that at one stage he got down on his knees, in what seemed a genuinely spontaneous gesture, and did obeisance unto us. (If you have seen Tony Bennett and know for a fact that he does this every time he performs, keep it to yourself. Even at sixty-one – especially at sixty-one – you need illusions to cling to.) Then he stood up, dusted the knees of a suit that probably cost more than my house, and finished his set.

Anita Kamath, my Radio 1 producer, and I were so moved by Bennett's performance that we tried to book him for the programme. He was going to do it, too, but had to cry off at the last minute with a bad throat. Fair enough, we thought, but we'd still like him to give us a song if the opportunity arises again. At least with Tony Bennett there were none of the problems we encountered recently with the American band Nile.

Nile play death metal, a form rarely essayed by Tony Bennett – although I'd like to hear him give it a go. Death metal is ideally very loud, very hard, very grim. Nile are, it must be said, uniquely good at it. They claim that some of their songs are sung in Sumerian, a language that has not, I understand, been heard in 2,000 years. On their 1998 album *Amongst the Catacombs of Nephren-Ka* are such teasingly Sumerian titles as 'Barra Edinazzu' and 'Kudurru Maqiu', and on the follow-up, *Black Seeds of Vengeance*, fans can sacrifice goats or be transformed, lo, into serpents which crawl upon their bellies, to such tunes as 'Nas Akhu Khan She En Asbiu' and 'Khetti Satha Shemsu'. They also suggest a high-risk strategy for dealing with war gods, should you encounter one.

When we tried to book Nile, someone in their organisation was more concerned with Nephren-Ka parking and when we were unable to guarantee space for a tour bus, the band pulled out. Tony Bennett's people never said anything about parking but if that's a problem, I'll go down to our Maida Vale studios and save a place for the singer any old time.

# Berlin Punk

*Radio Times*, 14–20 October 1995

---

HAVE YOU EVER heard the *schiffsbegrüssungsanlage*? Literally the ship's-salutation-unit, this is the device that plays *Deutschland Über Alles* to passing boats on the Elbe river between Hamburg and the open sea. You would think that people living in the houses grouped at the foot of the *schiffsbegrüssungsanlage* would have come to hate it long ago, but it seems not. Not only do some houses have hooters attached so that their occupants can join in the giddy fun of it all, but residents appear on riverside balconies to wave towels at the passing ferries. Whether they do this when a coal barge drifts by on the tide in the small hours I cannot tell. I have a feeling they might.

I heard the *schiffsbegrüssungsanlage* as I travelled to Berlin, via Hamburg, to take part in a film being made by a trio of women fronted by someone known only as Schneider. Roughly, the idea was that this film should follow up an earlier one about the lives of punks in East Berlin at the end of the seventies. As the main source of music for these punks was the programme I did – and still do – for British Forces Broadcasting Service under the memorable title *John Peel's Music*, it was deemed important that I should take part. Schneider and her partners turned out to be wonderful: impudent, imaginative, funny and still celebrating, as East German women, their double liberation.

Before we started work, they showed me the earlier film, made for television and the winner of several awards. This confirmed that being a punk in East Berlin was substantially more than a fashion statement. For a start, punkdom could bring with it a prison sentence for anti-social behaviour, as could writing letters to a British disc-jockey, the reason that most of the East Germans who corresponded with me did so when they were on holiday in marginally less oppressive countries such as Poland or Hungary. If you were a punk in East Germany, you were really a punk. No messing. Your attitude brought trouble, not only for you but for family and friends.

For women, it could also bring special unpleasantness in the form of a type of theoretical crime prevention that I hesitate to describe for fear the readers may be offended – or that it could be taken up by officials charged with waging war on single mothers. Women held to be liable to contribute to the causing of a breach of discipline (I know this reads like a clause of the Criminal Justice Bill but bear with me) were obliged to report to the authorities to provide a *hundegeruch* or 'dog-sniff'. This was a piece of cloth which women or girls – Schneider has just confirmed by phone that females as young as thirteen or fourteen could be compelled to provide a dog-sniff – had to rub beneath their arms and between their legs before officials sealed them in bottles and filed them away against the day when dogs might be used to track the women down. As an instrument of control and ritual humiliation, this takes some beating. As a practical means of crime prevention it would seem to be worthless.

Our filming was done on a pleasure boat on the River Spree. On board were some fifty or sixty punks from East Berlin. Many of these had, of course, moved on to lives far removed from punk, but others had been quite seriously disturbed by the treatment meted out to them as punks in the Soviet bloc and seemed still to be living the life as best they could, battling now with imaginary enemies, among whom at least one man listed Schneider and me.

Nevertheless, the cruise on the Spree was a great success. Much beer and food was consumed, and with conversation limited by my lack of German and the punks' unease with English (other than that used by the Sex Pistols in the television interview that ruined Bill Grundy's career), a deal of manly hugging and brotherly punching on the arm went on. A man called Muckel prodded me in the side with such frequency that I thought I would lose the use of my left arm. '*John Peel's Music*', Muckel would say, smiling wildly. 'Yes,' I would reply.

As we ate, drank, hugged, punched and prodded, the cameras rolled. The results will be on German television, well, some time. I came home feeling more European and less European all at the same time. With Schneider, her friends and most of the punks I felt entirely at ease. Indeed, I wish they were here now. On the other hand, there is the *schiffsbegrüssungsanlage*. That's plain weird.

# Bhangra

## Punjabi Homegrown at Noon

*Observer*, 28 June 1987

---

'RANJIT TO THE STAGE please,' urged the DJ, punching up the Nitro Deluxe record and thereby promoting substantial activity on the dance floor. Many of the dancers looked as though they should have been at school. Their customers ranged from the seemingly traditional, layers of brightly coloured materials arranged with such care as to suggest ceremonial use, to the enviably chic. Four wildly over-excited cake-walkers below me were clad as Beastie Girls.

A young woman, who had sidled up to enquire whether I was a reporter, explained that these discos have to occur in the afternoons as Asian parents will not permit their children, especially the girls, to go out after dark. Joe, saxophonist with Holle Holle, told me that, even so, many of those present would be there without their parents' knowledge.

Holle Holle are set to play at London's ICA on Saturday as part of a festival styled Punjabi Pop and the Bhangra Beat, subtitled 'A Five Day Celebration of this exotic homegrown phenomenon'. 'Homegrown' is fine – bhangra evolved out of traditional music played for weddings when, in 1977, Alaap became the first British-based Punjabi band to play what could be identified as a gig – but anyone going to the ICA in search of the merely exotic is due for disappointment. This is music for dancing yourself daft to.

In the recent past I have noted that a growing percentage of the names of those writing to me at Radio 1 have been Asian names and this has seemed to me, in some indefinable way, a good thing – the community, if you like, coming in from the cold. Having said that and knowing that the bhangra bands, like any others, want to play to as large and diverse an audience as possible, it is hard to avoid the conclusion that an event such as Wednesday afternoon's disco would be spoiled by any substantial European presence.

Virtually cigarette smokeless and with none of the sense of potential violence that mars some gigs – whilst, it must be admitted, enhancing others – this felt more like a wedding party. As the Executive Sounds disco, supporting Heera at the ICA on Friday, took over from Premi (who had played a blinder, establishing such a storm of rhythm around the heartbeat of the dholak drum that I yearn to hear them produced by the visionary Adrian Sherwood), I remarked on the presence on the floor of at least a dozen black women, suggesting to Joe that they might well be here because they could dance without interference from clumsy, would-be suitors.

Holle Holle's set was bedevilled by sound problems and the tiredness of the musicians. This was, explained singer Manjeet, former member of Alaap along with dholak man Chandu, their third gig in two days and, as Joe told me, all but two of the band have day jobs. Manjeet himself makes gearboxes for racing cars. Now, that *is* exotic.

Bhangra is still evolving fast within the Asian community and will flourish with or without the interest of the mass pop audience. A cassette by Heera apparently sold 30,000 copies in a week through corner shops, butchers and other non-chart return outlets. A lot of well-known pop bands would very much care for such sales. As for me, I shall go to further Punjabi discos, even if my mother forbids it.

# The Bhundu Boys

## Five Boys Go Big, Big, Big

*Observer*, 15 February 1987

---

IN SOME SMALL, well-nigh infinitely subtle way, I seem to have disturbed the even tenor of the Bhundu Boys' lives. Avid readers will recollect my enthusiasm for the Zimbabwean quintet when they – the quintet rather than the avid readers – played in London in May of last year. Since then their records have been prominently displayed on night-time Radio 1 and the band have played numerous gigs throughout Britain, the bulk of them in small rather than large venues.

Three weeks ago I saw the Bhundu Boys play in Chelsea and thought it one of the very best public performances I have ever witnessed; right up there with the Faces in Sunderland, Little Feat at the Rainbow, Captain Beefheart in Hollywood and Pink Floyd in Hyde Park.

In my charming, halting way, I endeavoured to communicate this enthusiasm to radio's teeming thousands, and herein, as I understand it, lies the germ of the complaint that Bhundu Boys' Biggie Tembo laid at my feet at the King's Head, Fulham, on Thursday night.

Biggie's point was that unstinting praise puts unwanted pressure on the band and leaves the novitiate to expect a transcendental experience every night. In the recent past we have been flogged the notion that African music is about to be big, big, big. To this end prominent African musicians have been signed to fabulous contracts, their work remixed by vogueish Frenchmen, their bands brought to Britain for prestige gigs.

By this process African music has been presented as an exotic novelty upon which the fashion-conscious can briefly alight before fluttering away to something else. The Africans have then been quietly abandoned.

The Bhundu Boys have done it very differently and, judging from the reaction of correspondents, much better. I have letter after letter (on display, upon request, in our offices) from everyday folk claiming that their reactions to the Bhundu Boys in concert have been similar to mine. Thus the Zimbabweans have, in a small way, but to greater and more lasting effect, entered the national consciousness.

With the LP *Shabini* high in the independent charts, major record companies are hot on the band's trail, but, judging from Biggie's remarks on Thursday, the Boys are unlikely to be swept off their feet by these attentions. In Fulham the Bhundu Boys seemed tired but still gave extreme pleasure. They play what is in effect superior pop music, catchy tunes with lyrics that, although they rarely stray into English, are still somehow memorable. For a demonstration of this latter, you should hear my performance of 'Hupenyu Hwangu'.

Playing a guitar loaned to him by Radio 1's Andy Kershaw, Biggie Tembo steered the Bhundu Boys through a ninety-minute set that rarely flagged and triggered off some serious dancing amongst a small and remarkably diverse audience. I do not want to risk further vexing the Bhundu Boys with excessive praise, but they do merit your most careful consideration.

# Bores

*Sounds*, 7 February 1976

---

THIS F. JONES of Cardiff person won't like it much, but I have just got to tell you about my rash. Certain sections of my body, principally my upper arms ('Like Mighty Oaks' – *Miami Tribune*), my fat, and my thighs ('You must be joking' – Noreen Vosper, aged fifteen, Bromley, Kent), are covered with patches of livid red.

I incline to the belief that this condition has been brought about by Liverpool's recent collapse at the hands of the bandit organisation known to lovers of Javanese wrestling as 'Derby County'. However, a medical person here at the BBC assures me that my complaint, which is accompanied by persistent and severe itching, is brought about by an allergy to some specific brand of soap or detergent.

I tell you all of this not to win your sympathy, but so that you'll understand what I'm doing if I unexpectedly break away from the typewriter with a loud cry and rush around the room cursing, clawing at my multi-hued flesh the while. I wouldn't want you sidling out of the room to alert the proper authorities over what is, after all, a passing condition. Now to business.

It was awfully good of some of you to vote for me in the Disc-Jockey section of the recent poll. I've dropped a few hints in the right quarters but it seems unlikely that I will receive either cup, trophy or illuminated address.

When I was a younger chap the papers were always filled with stories about worthies who were presented with illuminated addresses by a grateful community. I never – and I freely confess it – knew exactly what an illuminated address was. Reason told me that it must be something like '27, Rommel Villas, Doncaster' in neon tubing, but the same Reason then turned around and told me not to be so silly or I'd be sent to bed without my beetroot juice and *Health and Efficiency* annual.

But I digress – thank you, as I said, for your support. I think you picked the right man for the job now that Freeman is on the wrong side of fifty-five. But I do need to pick a bone with you, if you have a moment. I was terribly disappointed at being placed eighth in the Bore category. I have tried year after year to net this particular award and, just when victory seemed in my grasp, you plucked the victor's cup, as it were, from my lips. Or, to be strictly accurate, the Bay City Rollers did.

Second, naturally, was Timmy Bannockburn, and I was rather vexed that the judges refused to disqualify him from the competition. He seemed such an obvious choice somehow. I hear that some of the boys on one of these commercial radio stations have been making sport at the expense of Terry's version of 'White Cliffs Of Dover', and not only is this unpatriotic, but also they are mocking the afflicted which, as we all know, is quite wicked. Bread and water for a fortnight, pals.

Third Bore was Bruce Springsteen. Now, Bruce has certainly had more than his fair share of press in the past twelve-month, but I did get to see him in action in London and he really wasn't all that bad. Certainly nowhere near as awful as artfully down-and-out Tom Waits (and if you haven't heard of him then I very much fear that you will soon) and Patti Smith, who must be worth a quid of anyone's money as Bore of 1976. Try an each-way bet on her for 1977 while you're at it.

Fourth came Telly Savalas. A year ago I'd have joined you in voting for Telly, but since then I have actually sat down and watched the odd *Kojak* or two and I must say I rather enjoy them.

Fifth, sixth and seventh places went to Steve Harley, *Top of the Pops* and Queen. I never see *Top of the Pops* these days – my man in Harley Street has advised me against both it and *Celebrity Squares* – and I have a suspicion that the 'Bohemian Rhapsody's long run at the very tippy-top of the charts may well militate against Queen in the not too distant future. Certainly teenagers with whom I'm in contact (when their parents' backs are turned) in the quaint rural community in which I live tell me that they are heartily fed up with hearing the confounded thing. A joke, they feel, is a joke but really this has gone too far.

I had planned to nominate a few bores of my own – I mean, where was Michael Parkinson? Henry Kissinger? Richard Burton and Elizabeth Taylor? The last-named would, I think, have won themselves some sort of special award for being undescribably boring for at least a decade. Every time the *Daily Mirror* slithers through some gossip or solid news about this uniquely wearisome pair. Cannot the daily press include a detachable Burton/Taylor supplement so that those of us in danger of being pushed over the brink by their antics can chuck it firmly over our shoulders each morning?

I'd like to discuss this topic with you a trifle longer. Sadly I feel the need to adjourn to a private place, remove my clothing, and scrub myself down with the lavatory brush. The rash is flickering like a distant beacon, now red, now amber, now green. What I need is a tub of hot Calamine Lotion.

# Bottom Lines

## AC/DC in Concert

*Observer*, 19 January 1986

---

ON A RECENT TOUR of America, veteran rockers AC/DC were accused by the burghers of Springfield, Illinois, of being in league with Satan, and there is no doubt that their lyrics must win them a high ranking in feminist demonology, but the truth is that the band is essentially a well-drilled vaudeville act, hardly meriting such earnest attention.

I first saw AC/DC at a Reading Festival in the mid 1970s. The high point of their performance there came when guitarist Angus Young (with his brother Malcolm a survivor from the original band) showed us his rather unremarkable bottom. During the interval at Wembley, following a raucous performance by Fastway, I discussed this and related matters with a range of young men in denim, a fraction of quite easily the most genial audience encountered on my recent travels. The informed view was that we were likely to be denied Angus's bottom on this occasion. My confidants, in fact, seemed to know exactly what AC/DC would do, and in this, or so it appears, lies the secret behind the band's global success.

From the moment the band takes the stage it is clear that Angus is the star. As he wears velveteen shorts with matching jacket, white shirt and dinky white socks, cavorting about the huge, bare stage in a manner owing as much to Max Wall as it does to Chuck Berry, he would take some upstaging anyway. Singer Brian (Beano) Johnson comes close, devoting a disproportionate amount of his time to adjusting his costume in a meaningful manner, while the rhythm guitarist, bassist and drummer, who fulfil their roles exceedingly well, are manifestly of the other ranks.

'Angus, Angus,' the crowd carols, and Angus responds, tossing his hair about with a wit and panache unmatched since American guitarist Ted Nugent, a virtuoso hair tosser, at another Reading Festival. The guitar posturings, a vital ingredient, were not overlong and the volume, although intense, was not enough, as I had secretly hoped, to cause bleeding from the ears.

During a number possibly called 'Jailbreak', Angus Young performed a modest strip-tease, sparing us the bottom. The predominantly male audience goes wild. I am tempted to inquire whether any of them fancy him, but wiser counsels prevail. A similar inquiry made in the ladies toilet by my researcher produced only puzzlement.

This is medicine-show stuff at heart, a routine and predictable performance with every nuance anticipated and appreciated by an audience which resents change. From those portions of song that were discernible through the din the cardboard world of AC/DC becomes clear. It is a world in which women are forever sixteen, claim to be virgins, may transmit venereal diseases, and are eager to rock 'n' roll all night long. Ultimately to an outsider the whole thing becomes like a crowded and extremely good-natured masturbation ritual. As a reluctant voyeur, I tiptoed away before a climax was achieved.

# Butthole Surfers

*Observer*, 16 August 1987

---

CONFRONTATION STARTS EARLY at a Butthole Surfers performance. Outside the Clarendon, Hammersmith, on Thursday, a crowd of several hundred souls was attempting to push its way by main force through a single narrow doorway. 'My name's on the guest list,' cried many of the pushers hopelessly. As a member of the ticket-holding class I felt elitist and vulnerable. It took me over half-an-hour of heaving to get in.

The Butthole Surfers, who, as their name suggests, are not one of those bands hoping for a fun-run on the underground scene before exploding into big pop, have, during their skulk to a kind of ghastly prominence, driven rock writers to bulldozing adjectives into heaps. *Sounds*, for example, has described their music as 'subversive acid-crazed trash' and the band themselves as 'wigged-out surrealist purveyors of sonics'. Far from being overstatement, this colourful stuff only scratches the surface of a band apparently seeking a kind of purity through excess.

Once inside the Clarendon, where A. R. Kane are having their set terminated by an official with his eye on the clock, it became clear that some of my fellow-revellers were determined upon distancing themselves from reality at speed. Several young men fixed me, in the course of the evening, with a gimlet eye, murmured a considered 'Hi', and stood back for me to unstitch the mysteries of the universe for them. Their general appearance and choice of costume suggested that they had been rather carelessly dug up a week or so after an only partially successful embalming and burial.

The Texans appeared on stage as a film of police and ambulance crews removing dead Americans from highway accidents was projected at them and they launched without preamble upon an hour of music which made all known forms of disorientation technique seem pretty small beer. The widely dilated eye was taken initially by a woman dancer, first topless, shortly afterwards quite naked, whose frenetic activity quickly passed beyond eroticism to become, amid the smoke, occasional flames, bedlam guitars, flashing lights, pitiless dual drumming, film and slides, something little short of nightmarish.

Butthole vocals, which are rarely discernible and hardly need to be, emanate from the sizeable Gibby Haynes, whose father is a children's entertainer back home in Dallas, and, sometimes delivered through a bullhorn, serve only to heighten the sense of threat and disorientation.

When I stumbled out into the night, fat and sweaty and clutching a magazine I had been sold called, rather neatly, *Fat and Sweaty*, attitudes were in disarray, judgement lay bleeding and battered. Seeing the Butthole Surfers, whether they are visionaries or charlatans, is one of those experiences that compels you to reassess much of what goes on about and within you. I was disappointed at how few of my Radio 1 colleagues I spotted in the heavy throng.

# California

*Radio Times*, 23–29 March 2002

---

IT WAS NEARLY midnight. I had walked the dogs and was anointing my body with unguents in preparation for sleep when the phone rang. The call came at the end of a night of phone calls, one of those nights on which all you want to do in the entire world is to sit semi-conscious before the television and watch an hour or two of real people falling off things or out of things without hurting themselves too badly. 'We pay £250 for any film we use.' You know the sort of thing. Can't get enough of it.

The call came from Los Angeles. In November some of the world's greatest turntablists will be converging on the sprawling metropolis (I write all this stuff myself, you know) for a weekend of turntablism. The caller wanted me to join them. (For those of you uncertain about turntablism, it involves creating music on two, three or possibly more turntables from old records. It requires a sense of rhythm married to a keen ear, a keen eye and astonishingly agile fingers. Needless to say, I have none of these attributes.)

'But I can't do that sort of thing,' I admitted.

'Doesn't matter,' said the caller.

Sheila was already in bed. 'They want me to go to LA,' I said. 'To a turntablist weekend.'

'But you can't do that sort of thing,' she encouraged.

She didn't add, 'Why, you don't even play records properly on the radio. Wrong tracks at the wrong speed, that's you, that is.' But I bet she was thinking it.

I was last in California thirty-five years ago and would, I decided, like to see the state again before I grow too old to dream.

Also, given the speed of change there, it would be, for the locals, like being visited by a man who remembers the gold rush. I'll go, I thought, if only so that I can take the freeway out to San Bernardino and try to find the wooden house in an apple orchard in which I lived for eighteen months. I would predict that the house – little more than a shed, really – has, along with the orchard, almost certainly been bulldozed and replaced with a condo, but I've never been absolutely sure what a condo is.

I'd also, to be honest, quite like to drive slowly by the orange grove where I used to meet, for intermittent light snogging, Landa, one of those olive-skinned California girls I had previously imagined to have existed solely in the eponymous Beach Boys song, if only to reassure myself that I haven't always looked like a minicab driver down on his luck. I should apologise to minicab-driving readers here, but the truth is that whenever, dimpling modestly, I let slip the reference to the fact that I believe I look like such a driver, my audience, having moved initially in the direction of polite protest, pauses to look me up and down before replying, 'Actually, come to think of it, you do look like a minicab driver.'

At sixty-two – 'No, you're not! You can't be!' – I'm beginning to feel a powerful urge to return to some of the scenes of my early failures while I can. Under the circumstances, I'd better finish this column and start working on my turntablism.

# Captain Beefheart

## Yellow Brick Road Revisited

*Observer*, 6 April 1986

---

PAINTINGS BY THE American Don Van Vliet are on view at the Waddington Galleries in Cork Street, London W1, until 26 April. Van Vliet was born in Glendale, California, in 1941. Legend has it that he showed early signs of artistic ability but refused a scholarship, instead joining a band called the Blackouts.

In 1964 Don Van Vliet evolved into Captain Beefheart and formed the first Magic Band. Within a year the combo had established a substantial local following and was signed to A&M of Hollywood. The first release was a single, 'Diddy Wah Diddy', a Bo Diddley song, which I played as much as was commensurate with public order on radio station KMEN in San Bernardino, California. This enthusiasm was rewarded with an invitation to a live performance at the Whiskey-A-Go-Go in Hollywood.

Headlining on this most memorable night were Them, still fronted by Van Morrison, but they were as nothing in my eyes and ears when compared with Beefheart and the Magic Band. He sang with the voice of a desert Howling Wolf, his musicians squalling away behind him, producing a disassembled version of rhythm and blues, which, although it owed something to the Rolling Stones and their like, conjured up visions which had nothing of Chicago's South Side nor London's King's Road to them.

The follow up to 'Diddy Wah Diddy' was 'Moon Child', written and produced by David Gates, later leader of the indigestible Bread, but it was the LP *Safe as Milk* which introduced Beefheart to a European audience. I was so smitten with this LP that when the band first played in England, at the Middle Earth in Covent Garden, I burst into a torrent of tinkling, fairy tears as I introduced them on stage, subsequently hiring a car to drive Beefheart to his other engagements in Britain.

Captain Beefheart has, for twenty years, been the man at the heart of the music that has stirred and excited me, and echoes of his work, if not the work itself, are heard in every radio programme I do. I mentioned this to him as we stood in a light drizzle outside the Waddington Galleries on Wednesday. He laughed. 'The drums,' he said, 'they never get the drums right.' Then, when I looked puzzled, he laughed again. 'It's all right, I'm joking,' he said.

He was standing in Cork Street, clutching the iron railings, with his wife Jan, because he was ill at ease about facing the gallery-goers within. 'I was never like this before music things,' he claimed, with notable inaccuracy.

Over the years Captain Beefheart's reputation as an eccentric has been developed to an absurd degree. His jokes and word-plays, isolated from their natural context, are reproduced with great solemnity, with the consumer being invited to read profundity into observations which are often deliberately meaningless.

The real Don Van Vliet is a good-natured man, quick to laugh, with a deep mistrust of the record industry and an appetite for words and, on the evidence of his paintings, colour. Anyone who gives his songs such titles as 'Sun Zoom Spark', 'Big Eyed Beans From Venus' and 'When I See Mommy I Feel Like A Mummy' is not a man who wants us to take him totally seriously.

'When I See Mommy' is also the title given to one of Don Van Vliet's paintings, hanging alongside 'Red Shell Bats', which would make a fine LP title, and 'See Through Dog With Wheat Stack Skirt'. In my head I can hear Captain Beefheart singing that.

# Captain Beefheart 2

## Peel Meets the Captain

*Disc*, 1 April 1972

---

SO HERE I SIT typing while the Pig drives a Land Rover full of John Walters and Helen round to a neighbouring hostelry. The weather's broken and it's become cold and windy but I can still see and hear the lambs in the field across the lane and this morning's rain brought the level of water in the stream up an inch or two.

Last night we sat and watched the Eurovision song thing and tried to assess which song would win. All four of us did well although Pig and I both thought that the Austrian song, which was the only one that sounded as though it might have been written after 1950, would do better than it did. The song contest, like the Miss World show, is one of those things that you really have to watch because memory dulls during the year and you start to wonder if it really is as dreadful as you remember. Neither of them ever let you down.

After that we watched *Match of the Day* and John and I debated my theory that it has become an unwritten rule never to mention Liverpool on television or radio. With the team running fourth, stomping Manchester City 3–0, Everton 4–0 and Newcastle 5–0, you'd think that they would come in for some consideration – or at very least a mention in dispatches, but no.

Thursday was a big day this week. Pig and I drove up from the cottage, after I'd changed a flat tyre amid curses, for a reception at the Speakeasy for Don Van Vliet, Captain Beefheart. Three years ago, when the good Captain last came here, I abandoned most of my regular activities, hired a small car and drove him to as many of his gigs as I could. In those days he and the Magic Band were pretty much unknown quantities and yet the reactions of the audiences were strangely uniform.

At all of the performances I saw and heard, about half the audience would get up and leave, muttering or laughing, after about five/ten minutes, while the other half would stay and allow the Captain's music to flow over them and ultimately involve them. For the unwary he must have posed a lot of problems and I'm not really surprised that many people felt unequal to coping with the radical changes in taste and attitude that were required to absorb what he was doing.

He was then a very nervous and apprehensive man who saw potential enemies and villains lurking behind the most innocent of covers and, whilst enjoying the time I spent with him several years ago, I was apprehensive of meeting him again – especially at a reception. In the event the passage of time has wrought changes on him – changes I should have anticipated from the changes in his music – and he has become a happy and contented man. That was a joy to see.

Laughter is never far away and despite the certain knowledge that he's playing a different game to most of the rest of us, the feeling is that at least he's in sight now. He's still full of puns and jokes, which often make no sense until you think your way laboriously through his thought processes when you have the time to spare later, and can spin such webs around people who insist on asking him weighty questions that they are pulled to a standstill.

In Bristol he whistled the theme from 'More' when they called for 'More' and laughed for a minute when telling the tale over the phone. In the world of contemporary music just about anyone can be called 'genius' for what they can do with what has been done before. Captain Beefheart is out there charting his own erratic way through the heavens and must be the only real visionary to have brought his power to our music. We should treasure him.

The same night we went to the Albert Hall to see Leonard Cohen and that was a bit of a disappointment. Leonard Cohen's bleak vision had never seemed more fitting to me than it did when we were driving slowly through the Black Forest in a fine drizzle and in the florid Albert Hall it seemed inappropriate. The sound system was less than clever also and, whilst disagreeing with friends who feel that the blasted landscapes and echoing empty streets he conjures up are unhealthy, I didn't feel that the concert was a great success. Leonard Cohen is the sort of singer you should come across unannounced in some despairing place. Later he said he remembered me and we were at a reception in a Chinese restaurant which was attended by a torrent of smart and fashionable folk from smart and fashionable papers and TV programmes. I didn't see any other radio folk there at all. Later still when we passed him in the street he didn't remember at all but can hardly be blamed for that. Being lionised must be a tricky business.

The Pig and Gerry, who'd come with me, slept on the rocks on the Greek island on which Leonard Cohen once stayed and say he was a friendly and a happy man. He looked harassed and frustrated the other night. Perhaps he needs an application of what the Captain has found.

# Chicago House

*Observer*, 8 March 1987

---

IN THE BEGINNING there was Frankie Knuckles. Or rather, on this occasion, there wasn't. Frankie Knuckles is, by common consent, the father of House music – the name derives from Chicago's Warehouse Club – and he should have been setting the pace for the Chicago House party at London's Limelight on Thursday. The problem was that Frankie has no passport and nobody realised this until it was too late. Or so the story goes.

House has been staple fare on some of Britain's peachier dance floors for a couple of years now, and two House records, Steve Silk Hurley's 'Jack Your Body' and 'Love Can't Turn Around' by Farley Jackmaster Funk and Jessie Saunders, have brought lustre to our charts. Originating in Chicago's black, gay clubs, House is, roughly speaking, a fusion of seventies disco, new rhythm tracks and the sort of electronic noises associated with Kraftwerk.

House is often repetitive to the point where the passer-by might be forgiven for believing the record had stuck and is not, I venture, anywhere considered to be music for listening pleasure. Weary suburbanites will never return home and settle down to cocoa and Nitro Deluxe's 'Let's Get Brutal' or 'Hey Rocky' by Boris Badenough. Or if they do, I don't want them living near me.

Unfortunately, whoever it was who was staging the presentation of the House party, all Trax Records artists, had chosen to introduce the performers with airy music of the type you might expect to accompany a film called *Norway, Land of Enchantment*, and the spell was, for the dancing hordes, broken.

Someone who could see the stage – today's young people are too tall, I find – told me that the first performers, a glamorous duo who fidgeted with keyboards and jacked their bodies (danced) a lot, had been introduced by some dignitary swathed in bandages. I'm sorry I missed that. Despite the melancholy fact that the dancers seem to have been transformed, in a twinkling, into tourists, many with cameras, and the sound, which had been appropriately crisp and brutal for the records, gave the impression that it was being brought up from the sub-basement in buckets, it was clear that House is one of the areas where pop music's elusive energy is alive and well. House may only be, as someone sniffily howled in my ear, a passing fancy, but even if this is so, I cannot see that guaranteeing dancers a few years of exhilaration is a bad thing.

As the performance progressed it became clear that whatever House may be, it is not music for looking at. The artists, whose costumes relied to a marked degree on lurex and vinyl/leather, fought bravely on as the congregation treated them with the studied indifference accorded to buskers who may shortly ask for money.

# Children's TV

*Radio Times*, 9–15 December 2000

---

OK, SO I'M NOT A CHILD. And when I was a child, we didn't have TV. Therefore I'm not one of those ancients who murmurs, 'Ah, *Muffin the Mule*' (or *The Flowerpot Men*) and gazes wistfully out of the round window when the subject of TV for children is raised. I did have a brief *Magic Roundabout* period in the late sixties but, when it comes right down to it, I'm a *Clangers* man.

But having sired four children and watched them grow to something resembling adulthood, I've seen a lot of children's TV over the years and haven't much cared for it. Neither have the children, as far as I can recall. Well, William was made for the *Mr Men*, that's true. And Alexandra was solidly down with *My Little Pony* and *She-Ra*. Thomas used to watch *He-Man*, Sheila thinks, and Flossie cared deeply for *Button Moon*. All right, they watched a lot of TV. But what they didn't like was relentlessly perky presenters cajoling baffled children in brightly coloured studios, shrieking at us viewers and making the sort of terrible jokes they think listeners under the age of ten will find hilarious. So I don't think they would have liked *VIP* much.

I don't want you to go away thinking I'm a miserable old git, but I've never seen *VIP* before and I've also never seen Shauna Lowry and Robert and Elaine before. What's more, I'm not entirely convinced that the members of the Loughborough

Rugby Club under-nines team have either. They certainly look baffled when Shauna and the gang crew leap out at them and yell, 'You've been VIP'd!'

The under-nines, loads of boys and one girl, are coached by Adrian, and *VIP* people descend on them all out of the blue when they're training and tell them that they're going to have a medieval banquet. Then the children shriek, albeit in a dutiful sort of way. By a curious coincidence, Lynn, assistant producer on my Radio 1 programme, arrived as I sat down to watch *VIP* and she'd just been to a medieval banquet in a castle in South Wales and was able to give me a few pointers.

The banquet is to consist, Shauna shouts, of chicken, honey, oats and vegetables. The children groan, in the same way that children surely groan inwardly when over-sensitive parents give them trains made in Norway from sustainable hardwoods rather than Space Carnage 3. 'Look, Ryan. You can put all the little cross-country skiers into the carriages. Don't lose their rifles dear, will you?' 'No mead for them, then,' Lynn muttered, her eyes glinting.

'You gotta get with us, 'cos we got it goin' on,' a record blares on the soundtrack – that's the way young people talk in Loughborough, I expect – as Shauna and Elaine help the children to make appropriate banners and Robert helps Adrian the coach to make oatcakes with medieval oats and medieval flour and medieval everything you need to make Ye Olde Oatcakes, including, reason insists, medieval patience.

Then everyone speeds away in *VIP* cars to the World of Robin Hood (and the Sheriff of Nottingham and Little John, who 'fought Robin Hood on a bridge just like this one'), where Will Carling appears as if by magic in front of the children. 'Ex-England rugby captain,' trills Shauna, in case they've missed the point. Will copes well, as far as we can see, and answers questions that, in my opinion, the children have been prompted to ask him. I wouldn't last five minutes with a bunch of eight-year-olds.

When we had parties at our house, I would retire early in the proceedings to listen to beat music in my room, emerging only if civil insurrection threatened.

Another surprise guest is Otis from *Live and Kicking*. I could have sworn some of the pocket-sized rugby players knew who he was. Otis is dressed as a minstrel. You know, a medieval minstrel. More contemporary minstrels come in the form of Daphne and Celeste. 'Pop duo Daphne and Celeste,' urges Shauna helpfully. The pop duo mime to 'School's Out' with some of their friends. 'This much like your medieval banquet?' I asked Lynn. It turned out it wasn't. Lynn was dressed as a nymph, she admitted, and was obliged to go a-questing hither and yon in her Welsh castle. She also slew a dragon and attempted to slay Merlin the Magician using spells and potions (M&Ms and Tic Tacs apparently). She had a wonderful time, even though the train she was told to take didn't stop at Chepstow after all and went on to Gloucester and she was two hours late. So, I don't doubt, did the Loughborough Rugby Club under-nines. Have a wonderful time, I mean, not go to Gloucester. Oh wow. Fantastic, as Shauna would say.

Oddly enough, I was at a party myself over the weekend, one at which several top showbiz figures, including Angus Deayton and Ian Hislop, were guests. There were fireworks and it was raining, and in an attempt to appear an amusing fellow I offered Angus my raincoat, suggesting that as a city boy he might find the rain not to his liking.

'Thank you, I have a coat,' he said politely, clearly under the impression I was a demented rustic, probably one with a job connected with drainage. Sheila sighed and pulled me away.

# Kenny Dalglish

*Guardian*, 12 August 1994

---

I HAVE TO ADMIT we have stopped celebrating Dalglishmas since Kenny went to Blackburn. We never did anything too elaborate on his birthday – no bonfires on the village green or chestnuts roasting on an open fire – but we did open a bottle of champagne-style drink and toast the great man.

Of course, we were well aware of Kenny's abilities long before he came to Anfield. I stood on the Kop for his first home game and we warmed to him in a way that we had never really warmed to Keegan. The most impressive of his abilities in those early matches was the way he seemed to know where every player was at all times. Even a superb Liverpool squad, by far the greatest team the world has ever seen, took a few weeks to catch up with Kenny's speed and anticipation.

As with most footballers, Kenny's musical taste is deeply suspect, as I found when he came into the Radio 1 studios for a programme called *My Top Ten*. Or was it Twelve? I popped into the studio as the final notes of some Lionel Ritchie horror faded into nothingness and introduced myself. I had known for some time that I was going to do this and had slept poorly the night before, fearful that Dalglish would be hostile, stupid or both. Sheila had persuaded me that he would be embarrassed if I took the children into the studio for him to bless them. 'Not even

Thomas James Dalglish?' (our third-born) I asked. 'No,' she said, gently but firmly. Kenny turned out to have a neat, dry sense of humour and gave me his home phone number in case I ever needed tickets. Sometimes I did. I called him for European Cup Final tickets when Liverpool were to play Juventus. Sheila and I picked them up from him at the team's hotel on the day of the final. Kenny was in high spirits and we went off to the Heysel Stadium rejoicing that Juve had no chance . . .

Dalglish turned out during a pre-season match in Dundee last week and we wish we had been there. We send him a family Christmas card each year, although we never get one back, and conversation comes to a standstill in our house if Kenny appears on TV. We have the Dalglish *This Is Your Life* on video, although we haven't watched it for a while. We scrounged tickets when Blackburn played at Portman Road last season and Sheila and Tom sat right behind him. They didn't dare speak because Blackburn lost. Neither did I.

We all need heroes. Kenny Dalglish is ours. Still.

# Devil's Music

*Radio Times*, 29 July – 4 August 1995

---

'I DON'T EVEN REMEMBER which year it was. I was driving the '58 Biscayne from Dallas to Boston, the wind through the open window tugging playfully at my golden ringlets. In the back of the car slept a mother and child, neither of them much to do with me.

The mother had written one too many hot cheques in Texas and had opted for interstate flight to avoid prosecution. It was raining. It was about four o'clock in the morning. There wasn't much on the radio so I dialled up another station, as much for the exercise as in the hope of finding anything worthwhile. 'Dry bones in the valley,' a voice said, hitting the 'valley' hard and drawing the 'ey' out to breaking point. Then it said it again. And again. Behind the voice a congregation murmured in approval. I had found the Revd Mr Franklin, father of Aretha and Pastor of the New Bethel Baptist Church in Detroit, Michigan, and he was a few minutes into what turned out to be a twenty-four-minute sermon.

As the Revd Mr Franklin warmed to his theme, he roared, he sobbed, he growled. A deep, dark and unearthly throb came into his voice as he testified, and there in the middle of Illinois, in the night rain, I came as close as I ever have to becoming a believer.

Settling back to watch Scottish Television's *The Rock That*

*Doesn't Roll,* billed as an exploration of the expression of Christianity through music, I thought of the Revd Mr Franklin, his daughter and the hosts of black gospel and gospel-derived singers whose records I treasure. In programme one of *TRTDR* I didn't get them. Instead I got the Salvation Army. Now, I'm a bit of an admirer of the Salvation Army, two of its foot soldiers having saved me from a certain amount of unpleasantness when I was a troubled young gunner fighting for freedom and democracy by painting trucks on Salisbury Plain. So I bore with presenter Tom Morton as he introduced ancient film of the Army on parade, bringing a banging tune and a measure of practical help to the bruised and baffled.

I even coped with Tom's own picking and singing, despite the baseball cap. But my goodness and generosity of spirit evaporated with the Joystrings. The 'strings, some may recall, were the Salvation Army's answer to the Beatles. They had as much in common with the Beatles as I do with Radovan Karadzic, of course. Lt Col. Sylvia Dalziel, a former Joystring herself, told us that the combo was accused by other soldiers of the Lord of playing the Devil's music. If all that limp, po-faced stuff with its twittering and bleating, its greeting-card levels of spirituality and its bloodless strummity-strummity-strum is the Devil's music, well, it serves the blighter right.

What is it about religion that makes intelligent people content with Today's Lovely Thought-style expressions of it? I'd prefer something dark and medieval, all guttering candles, wordless mutterings in unlit corners and being zapped for coveting your neighbour's handmaiden.

Programme one of *The Rock That Doesn't Roll* (what does that mean, exactly?) ended with Tom Morton, looking significantly better without the baseball cap, in Memphis, Tennessee, and therefore several steps closer to the spirit of Aretha Franklin and her dad.

*

Typically, I went to the first Isle of Wight Festival, the one no one remembers (who, apart from Jefferson Airplane, played there?) and missed the second, at which French anarchists – and where are they, now that we need them? – tore down the fences and declared a free festival, and at which Jimi Hendrix played. Having seen Hendrix live a dozen times or more, I have since claimed that he was the greatest live performer I have seen. Having watched *Jimi Hendrix Plays the Isle of Wight*, I'm not so sure now. Alison Howe, producer of my Radio 1 outings, was certain. 'I just can't take this seriously,' she said. Perhaps you could argue that, as the Joystrings unleashed numberless hordes of drip-dry gospellers upon us, Hendrix unleashed legions of posturing, pouting guitar abusers on the world, ninnies whose antics have become so predictable and just plain silly that when we see the original again, he looks like another one of them.

It is, I decided, Hendrix's playing within the songs that stands out and is worthy of your attention, rather than the mildly embarrassing solos. Old hippies who danced or sat on the floor pretending to be stoned at London's Roundhouse will be pleased to see DJ Jeff Dexter introducing Hendrix.

# Doomsday

*Radio Times*, 23–29 May 1998

---

HOLD ON THERE, mate. You got a minute? You did know that all that Deirdre Barlow stuff was forecast in the Book of Revelations, didn't you? No, really. I don't remember the exact verse or anything, but it was. And Nostradamus wrote something about a red ball in the southern sky. Has to be the Arsenal's championship triumph, doesn't it? Obvious when you think about it.

One thing I am prepared to prophesy myself is that we're going to get a lot more of the above downloaded into our lives as the Millennium draws – as we say in the prophecy game – nigh. But 2,000 years of what exactly? And starting when? And why? That's what I want to know.

Now, I'm a bit of a pessimist. I admit that. I wake every morning rather expecting to be struck by an asteroid or for extra-terrestrials to come marching right down our road, right past Flossie's friend Beck's place, right past the Brigadier's and right into our garden. (Actually, it would be left into our garden but you know what I mean.) Once settled, the aliens would start with their death rays, eugenics programmes and unearthly sort of whistling noises. When these things don't happen, I crawl into bed feeling pretty pleased with the day, even slightly smug.

Your optimist, on the other hand, bounds from his or her bed believing that world peace will shortly be busting out all over, that a cure for just about everything is going to be found and, hey, it grows right there in the rockery and that, one day soon, all eleven members of a Liverpool team are going to play to their potential at the same time and in the same place. At the end of a day in which none of these things happens, your optimist, surely, must feel a bit crestfallen.

So, it was with a bit of a there-I-told-you-so on my lips that I watched ITV's *Knocking at Doomsday's Door*; the only reason that I didn't come right out and say it being that I was by myself in a hotel room in Bradford.

There have been loads of forecasts of imminent doom over the years and, so far, they've all come to nothing. But there's no doubt that asteroids, famine, floods and pestilence are all barking at the gate, anxious to be at us. Take floods. Take Lori Adaile Toye's maps of the post-apocalypse world. Lori Adaile got the maps from some boys called the Ascended Masters who, one suspects, probably offered to tarmac the driveway of her 'I am America' Centre in Arizona while they were in the area.

I love living where we do, but I have often expressed the wish that the sea wasn't the best part of 50 miles away. Well, according to the Ascended Masters' maps it won't be. In fact, we'll be several fathoms beneath it, which makes fussing about the damp patch in the hall rather pointless.

And asteroids. It seems there are hundreds, thousands, even tens of thousands out there, patiently queuing up to zap us. Even as we speak, men and women of the type known as boffins are peering into the skies and discussing, in rather attractive Middle European accents, total obliteration.

Then there's Antonio Carducci right there in Texas. Does Antonio know something we don't know? He seems to, right enough. He's gearing up for the End Time or the Tribulation or the Something, having used up a fortune on building concrete domes, storing tons of tinned food and treating himself to what looks like a mighty fine wig. You'll need to look your best on that great getting-up morning, people. Antonio is expecting hordes of children to descend on his concrete bunkers when the time comes. I forget exactly why. He has, bless him, made up a few extra beds ready for them. I haven't done my sums well, I admit, but I reckon on about 22,000,000 children per bed. The way to avoid this administrative nightmare, the ex-millionaire believes, is for us all to start being nicer to one another. Right now, I'm all for that.

# Driving to Sonar

*Radio Times*, 7–13 July 2001

---

THIS WEEK'S MYSTERY word is Ozanam. O-Z-A-N-A-M. 'Could you,' I had enquired at the information desk, smiling sweetly, 'tell me the name of the school the French children on board attend?' My smile widened. I must have looked wolfish. 'Why do you want to know?' the woman behind the counter countered. 'I'd love to write to the head teacher,' I explained, sort of. 'Well, all I've got on the log is Ozanam,' she admitted, warily.

We were on the ferry from Caen to Portsmouth. Sheila and I had left Barcelona after four days of Gaudí, electronic music, Picasso and meals with top dance DJs. I had contributed a ninety-minute set to the musical feast myself. I wouldn't want to appear immodest but it went down pretty well. (Advertising feature)

We had taken four days to drive to Barcelona's Sonar Festival. We spent the second night at the Château de Castel Novel in Varetz. We are not natural château people. The hotel staff were as crisp as lettuces: polished, cool, sophisticated, beautifully dressed. Sheila and I felt like charcoal-burners from the estate who had been invited up to the château for drinks. Perhaps, in the morning, they would perform hideous operations from beyond the boundaries of natural science on our drugged bodies. My last words to Sheila would be, 'They must have slipped something into the d'Yquem.'

As it turned out, our only real problem lay in tipping these demigods. Anything less than a million francs each would have been an insult but we felt they'd prefer a Picasso sketch, a fragment from a horn concerto in Mozart's own hand, an unpublished trifle from Byron. Search as we might, we could find no such thing so left nothing – as so many other guests must have done.

From Varetz we headed into Andorra in search of a more proletarian experience. We found it in Escalades which is a top destination for those anxious to buy watches, trainers or perfume. In a ten-minute walk through town we counted sixty-seven shops selling watches and almost as many selling trainers or perfume. Some sold trainers and watches. Others sold perfume and trainers. The occasional visionary sold trainers, perfume and watches. Actually, I bought some trainers in Andorra myself. You felt you had to.

We had less time for the journey home from Barcelona, spending a night in Cognac and a night in Caen itself. Having driven so far, so fast, and having had to get up at 5.45 a.m. to catch the ferry, all Sheila and I wanted once onboard was to sleep. So we found the reclining seats and settled in a corner. Then the schoolchildren arrived. They were hateful. Not just noisy but aggressive. They drove everyone but Sheila and me from the room. 'Remember the Glorious Gloucesters on the Imjin River,' I whispered. When we finally left for the car, the *garçons et filles* chorused a sarcastic 'Goodbye'. One of the two words I used in reply is unsuitable for use in this publication.

# Eastern Bloc

*Independent on Sunday*, 13 May 1990

---

AMID THE THRILLS AND SPILLS, alarums and excursions of the collapse of the dictatorship of the proletariat, the name of Wladyslaw Kozdra is little remembered. But it was this Polish party functionary who in 1978 strove to reassure his comrades that 'the trumpets of the Beatles are not the trumpets of Jericho which will cause the walls of socialism to come tumbling down'. Alas for the walls of socialism, the Beatles have proven to be better trumpeters than poor Wladyslaw a prophet for, as is demonstrated in Timothy Ryback's book, *Rock Around the Bloc* (subtitled 'A History of Rock Music in Eastern Europe and the Soviet Union'), the Beatles, with their predecessors and successors, have played a substantial part in undermining each of the former communist dictatorships.

Ryback details the clumsy and brutal attempts of the party to come to terms first with jazz, then with such phenomena as the twist, Beatlemania, rock and punk. Authority confronted the problems posed by youthful – and not so youthful – citizens inflamed by rock 'n' roll in the manner of a deranged motorist making progress by first accelerating wildly, then stamping on the brakes. Spokesmen hit out with speeches hewn from the living blancmange. 'Scurrilous ditties from the criminal world' and 'toilet stall poetry' were, they insisted, all part of a knavish Western plot to undermine communism.

If we had been that good at plotting, Romania, Czechoslovakia, Poland, Hungary, Bulgaria and East Germany, along with the Baltic republics, might not have suffered as they have, and guitar-toting guerrillas might indeed have been able, as a Russian book attacking jazz had suggested in the fifties, to 'deafen the ears of the marshallised world by means of epileptic, loudmouthed compositions'. Another ashen-faced pamphleteer warned that exposure to the indecencies of Western pop would lead inevitably to a dizzying descent from purity on 'a mudslide of boogie-woogie'.

As a man who has been content for years to wallow on such a mudslide, I have been considerably stirred in recent months by letters reaching me from Eastern Europe. After expressing enthusiasm for some of the bands whose records they have heard on the World Service of the BBC, most of my correspondents have gone on to recommend local bands whose music they believe merits a wider audience. The melancholy truth is that, on available evidence, they are likely to be wrong.

A few nights ago I stood, glowing with perspiration, in an upper room at the University of London Union, enjoying the antics of the University Stage Diving Team and the music of My Bloody Valentine. Where, I wondered, were the Warsaw Pact's My Bloody Valentines? And what would they make of this band, which coaxes from the music weeklies a torrent of adjectives of the 'shimmering', 'squalling', 'howling' class, in Gdansk?

In truth, we in the West have not been generous with the music we have sent eastwards. The Rolling Stones played Warsaw in 1967, but otherwise we have fed them such dull fare as the Tremeloes, Paul Anka, Boney M and Billy Joel. It is a measure of the craving for Western music among the Iron Curtain posse that they embraced even these anodyne offerings with reckless enthusiasm.

Worse, the traffic the other way has been, of necessity, not far short of non-existent. Music made within Eastern Europe has been entirely for domestic consumption and of symbolic at least as much as of musical value. Music and musicians have demonstrated to cowed populations that courage, improvisation and an ability to wrongfoot lumbering authority make an alternative way of life not impossible. To judge this music by our critical standards is pretty much a complete waste of time.

Last year, for example, the respected Russian band Zvuki Mu released an LP in Britain, came here to promote it and recorded for Radio 1 FM. Both the record and the BBC sessions were intriguing rather than inspiring. A concert I attended in Moscow two years ago, featuring Zvuki Mu and three lesser known local groups, was frankly dull.

There were far too many musicians for my taste, sporting foolish costumes, dressing as clowns or, worse, as monks, and behaving in a manner familiar to one who endured arts-lab theatricals in the late sixties. But the assembled Muscovites loved it, cheering lustily even in mid-song at encoded messages in the lyrics which I was incapable of detecting.

Ryback may find it necessary to update his book, which cannot be described as anything other than authoritative – although, incidentally, John Mayall was never a member of Cream – annually, if not monthly, as the newly liberated countries evolve. Many of the people who have written to me from the East have expressed their concern at what they see as the cultural colonisation of their finely balanced societies by a rapacious West, by traffickers raring to flog them such civilising devices as Garfields, vaginal deodorants and of course pop records.

But with luck there will be such a frenzy of activity in the basement clubs of Eastern Europe and the Soviet Union that some of the traffic will flow the other way.

In the meantime, can anyone teach me some of the dances devised by jittery commissars to supplant depraved imports like the twist, walk and Madison? Can you show me the lipsi, developed in Leipzig in 1958 and praised for its socialist qualities? Or the 1961 Soviet floorfillers, the progulka and the druzuba? Replies will be treated in the strictest confidence.

# Duane Eddy

## More New Numbers, Less Old Hat

*Independent on Sunday*, 8 April 1990

---

FROM TIME TO TIME in these pages, I hope to bring you the odd guiding principle. Follow them and your lives will be enriched. The first of these, arrived at after years of research, is simply this – There Is Nothing To Be Gained By Listening To Musicians Who Wear Hats On Stage.

I mention this because in his current promotional photographs, Duane Eddy is wearing a very serious hat indeed. I cannot confirm the guitarist wears this on stage but I have an awful feeling he might. He certainly wore the same hat – or its twin – on the cover of his 1987 LP for Capitol.

Cast your minds back, if you will, to 1987. The news was that Duane Eddy, whose name was synonymous with twangy guitar instrumentals – he even recorded an LP called, implausibly, *The Twang's the Thang* – was back in the studio. This was good. Further bulletins revealed that George Harrison, Paul McCartney, Ry Cooder, James Burton, Steve Cropper and David Lindley were in there with him. This was bad. When the LP was released it sold well but since first putting on my metaphorical wellingtons for a tentative tiptoe through the celebs, I have never listened to it again.

Now it is April 1990 and Duane is in Britain as part of the All American Solid Silver 60s UK Tour, replacing Del Shannon who sadly died by his own hand earlier this year. Also billed are The Crickets, who were unexpectedly enjoyable when I caught them a couple of years ago, and Tommy Roe.

Now I am not one of those Duane Eddy fans who is happy to hear him twang out the oldies all night long and, according to the publicity material for the tour, Duane himself has said: 'I want to go forward and be current and successful now.'

It depends how you define 'successful' of course, but Duane could take the example of another twangy guitar operative, Link Wray, to heart. Link is the American backwoodsman whose 'Rumble', according to legends half as old as time itself, persuaded Pete Townshend of The Who to take to the guitar. In the 1970s Link, hitherto a guitar/bass/drums sort of chap, was encouraged to sing and eventually get into guitar heroics of the most morbid type. The recorded results were pretty beastly and I had Link marked down as one of those artists who do not themselves fully understand what it is they do that is so good. But in 1989 someone from Ace Records persuaded Link Wray into Pathway Studios, London, to record tracks for an LP himself and left the vocal mike locked in the cupboard for all but two of the ten tracks. The result should bring a healthy glow to the cheeks of anyone interested in the survival of the rock 'n' roll instrumental. 'Apache' is by no means flawless – but who wants flawless?

'Flawless' means Eric Clapton.

I have never met Duane Eddy but he remains one of my greatest heroes. At times of stress our home throbs to his 'Hard Times', 'Movin' 'n' Groovin' or 'Three-30-Blues', and if our paths were to cross, I would urge him to imitate the action of Link Wray, hire a studio, find a sympathetic drummer, bassist and tenor player, write some new tunes and tell his famous friends that if they come within a hundred yards of the studio there will be a

Rottweiler in their futures. Duane and the rest of the tourists are in Nottingham tonight and they travel England and Scotland all month, finishing up at the London Palladium on the 29th. Leave the hat in your hotel room, Duane.

# Europe
## Suspiciously Like Handel

*Observer*, 22 February 1987

---

HOLD ON A MINUTE. There must be one in here somewhere. Yes, here it is: page 2 – the picture of the drummer sticking drumsticks up his nose. Now, let's see if we can find some reference to members of the band being 'classically trained'.

This is Europe's Final Countdown World Tour programme (£4) and William (11) and I are at the Odeon, Birmingham, waiting for the net curtain, rich with planetary motifs, to rise.

At 8.45 there is a considerable explosion, the curtain does rise and Europe are with us, playing 'The Final Countdown'. Rather plucky, we thought, to start with your biggest hit. Singer Joey Tempest shows early skills with the lightweight throwing microphone-stand and smiles a lot. Joey has a thoroughly winning, West Coast-style smile, a tribute to Swedish dentistry. Europe smile rather than frown, their music is loud but not too loud, their trousers bulge discreetly and believably and there is no trace of the Sex Criminal From Outer Space imagery usually associated with heavy rock.

There is much tossing of hair. Joey is especially useful with his hair and during the obligatory tormented guitar outbreaks he regards guitarist Kee Marcello with the conventional mixture of locker-room affection and superstitious awe.

Europe are aware of rock conventions but are not enslaved by them. Why, they even include a brief a cappella number and an extended guitar feature, which has much agile playing and quotes from (I think) 'The Flight of the Bumble Bee', yet boasted none of the half-baked histrionics one might have expected. A drum solo, restricted to a modest five minutes, started conventionally enough but became surprising when Ian Haugland left his kit and performed as drummer and cheerleader on a single snare set at centre stage. Europe's music is music of few surprises where normally there are none.

Experience has taught that it is usually keyboard operatives who are 'classically trained'. These artistes tend to sullenness – perhaps because it is in the nature of things that they cannot be as exhibitionist as their comrades – but Europe's Mic Michaeli seems content enough, his feature including outbreaks of what sounds suspiciously like Handel.

There is, of course, little difference between the songs. These tend to have vague but encouraging titles such as 'The Time Has Come' and all have what I cannot avoid calling the anthemic quality that made 'The Final Countdown' such a hit. At William's recent birthday disco I was compelled, during a set which otherwise failed to give total satisfaction, to play 'The Final Countdown' six or seven times. Europe's wholesome racket is metal for moppets, if you like. 'So young, so tough, so wild' runs part of one of the lyrics – but not *too* tough or *too* wild. Europe should last as long as their looks and ability to write songs that sound important last.

Towards the close, a supernumerary enters bearing a ringing telephone. Joey feigns surprise, answers it and talks briefly with a Swedish voice complaining about the noise. Turning to us, Joey says 'We have to stop – or do ya wanna rock the night?'

As we bellowed our affirmation, Europe slam into their current hit single. They rock, in fact, for four minutes.

# Eurovision

## Backstage at the Eurovision Song Contest

*Observer*, 17 May 1987

---

'THANK YOU,' said the King of the Belgians, 'for bringing joy to our country.' We of the British delegation to the thirty-second Eurovision Song Contest dimpled prettily as His Majesty moved away to meet and greet the Italians on our flank.

Since 1968, when Massiel won for Spain with 'La La La', I have had a well-nigh ungovernable appetite for the Song Contest, an enthusiasm fired by the unchanging rituals of the competition, the increasingly anachronistic songs themselves, and the blatant tactical voting of the national juries. Despite the latter, there is a curious innocence to Eurovision, an innocence that becomes, perhaps surprisingly, more evident when you witness the competition at first hand. The world of Eurosong is far removed from the cynicism and oafishness of rock, and wandering backstage among the contestants it is difficult to picture them trashing hotel rooms or scuffling after designer drugs. Kate Budd, a former *Opportunity Knocks* winner, singing backing vocals in Brussels, was able to say 'I'm going to give my all for Britain' and mean it, her transparent sincerity disarming criticism.

For Rikki Peebles, writer and singer of the British entry, 'Only the Light', winning would give him the chance, after years of frustration, to become a genuine contender. For Johnny Logan of Ireland, success would give him a chance to compensate for the disastrous career moves he made after winning the Song Contest in 1980.

For the others, representing their countries was generally regarded an honour, the contest itself taken seriously, clearly still of importance. In Britain Eurovision has become something of a fringe activity, its diminished status demonstrated by Radio 1's failure to add 'Only the Light' to its playlist.

For the noncombatants the week leading up to the competition offered a punishing round of parties and receptions, the most prestigious of which was the 'Soirée Saban', an entertainment organised by Belgium's music publishers and burned into my memory as 'The Night of the Twenty-two Belgians'. Taken in a procession of coaches to a restaurant on the outskirts of Brussels, we were fed and watered and treated to live performances by twenty-two of Belgium's most beguiling pop artists. As we left, having witnessed François Glorieux, Timothy, Pierre Rapsat, Les Gangsters d'Amour and the rest in intoxicating flight, each of us was presented by one of the co-sponsors with a box of tissues.

Prior to driving to Brussels, I had contrived not to hear any of the competing songs, preferring to arrive in the Parc des Expositions in a state of grace. Having sat through the first of three full dress rehearsals I noised abroad the view that 'Only the Light' was blessed with the anthemic qualities that, given a more portentous production, in the Ultravox manner, or larded with heavy metal trimmings of the type popularised by Europe, would guarantee chart honours.

Although the Danish entry, a rather simple-minded piece called 'En Lille Melodi', led the betting, it was generally assumed, after this first rehearsal, that all Johnny Logan had to do to win was turn up. My own preference was for the Yugoslavian song, a jaunty fifties-ish confection with a sort of hiccup effect in the chorus, sung by a disturbingly vivacious blonde woman supported by what looked like a quartet of border guards.

As a man who has sworn a terrible oath never to miss the Song Contest on television, I was delighted that it was still possible for me to see, in a hall where but a week before Boris Becker and the rest had been playing tennis, the short films which occur between songs to mask the movement of equipment on stage. This year these, apparently made by film-makers seconded from the picture-postcard trade, depicted every aspect of Belgian life, including, alas too briefly, dentistry.

Between rehearsals the VIP village backstage was cacophonous with hairdressing demonstrations and rotten with rumour. France's Christine Minier had fled from the stage in tears. The Israeli performance, inspired by the Blues Brothers and requiring much agility to present and a sense of humour to appreciate, had been denounced in the Knesset. Radio 2 commentator Ray Moore had offended the Turks to diplomatic level by remarking on the villainous appearance of their representative. Generally, though, it was all rather dull.

After the competition the winner, Johnny Logan, beside himself with excitement, made a startling speech in which he gave credit to Jehovah, denied that he was homosexual and spoke briefly in Turkish. This last had an especial poignancy as the Turks, having survived Ray Moore, failed to score a single point. The Greeks, on the other hand, came 10th out of 22, their score bolstered with a 12-point maximum awarded by Cyprus. The Cypriots themselves were wildly overexcited about finishing 7th, having, in turn, received 12 points from the Greek jury.

Rikki Peebles, in 13th place for Britain, a position which reflected, German friends suggested, a measure of Britain's unpopularity in Europe rather than any lack of merit in song or performance, put a brave face on it but was plainly heartbroken. As we left we were again presented with boxes of tissues.

# Eurovision 2
## Public and Private Taste

*The Listener*, 22 April 1976

EVERYBODY, BUT EVERYBODY, likes a good car-crash. Not much fun being a principal in the action, to be sure, but the stories one can tell if one just happens to be passing and, oh, my dear, there was blood everywhere and you should have heard the noise his girlfriend was making. Of course – hands raised in horror – we wouldn't like to see anyone really hurt. Not at all. The trouble with the car-crash is that it is over in a split second, and you can so easily miss the action. The motor accident is ripe for consumerisation, and if there is no such word as 'consumerisation', then I am very surprised. Advertise widely, slow the action right down, and it might just be worthwhile packing a few sandwiches and taking the kiddiewinks along. But until technology masters the accident and makes it marketable – and it will – we will have to settle for the Eurovision Song Contest.

I speak as a man who has sat without flinching through the last five of these events, viewing the proceedings in the privacy of my own home, and you'll have to take my word for it that the televised competition has much in common with the two-hour, head-on collision. There's the pacing, for a start; the terrible snail's-pace at which the head hits the windscreen, the gruesome build-up of entirely useless information about songwriters, band-leaders, artists. See in fine detail the inch-by-inch splintering of the screen, the films of the contestants in their natural habitats, frolicking, frolicking. Then, at the very end, there are the casualties. I think I shall never forget the slack-jawed and mirthless smile on the face of the spangled woman who sang Norway's entry, 'Mata Hari', as national jury after national jury ignored her efforts, to scatter their votes elsewhere.

The major difference between Eurosing and the motor accident is that no one emerges victorious from the latter. I have been wondering a lot since 3 April whether I should have felt a surging of my patriotic juices when the Brotherhood of Man won through for Britain. Will, I have asked myself, The Triumph In Which We All Share affect the sinking pound? Well, it might. If it is true that 450,000,000 persons teleview the Eurovision Song Contest, it might indeed. Contemplate, if you would, those 450,000,000. If – and I hope it never happens – you were to be suspended over Oxford Circus for eighty years, without sleep, that figure represents over three times the number of people who would pass below you during your rather pointless vigil. To be brutally frank, I made that last bit up, but the fact that it seems plausible is impressive enough. Viewed from just about anywhere, 450,000,000 is one hell of a market. And what sort of records do these 450,000,000 souls buy?

Over the past month, I have been compelled to consider the current state of popular music for a Radio 3 series called, to my regret, *Where It's At*. The six programmes in the series will be surging out on 464 metres medium wave, starting on 23 April. In such a series, and particularly in one with such a title, the listener may well expect to learn which are the current trends and in what directions we may find ourselves moving in the future. The first programme has just been recorded, and I can still, despite many hours of contemplation, see no trends or directions with which I can reasonably satisfy the questor after truth. Glib though it may be to say it, the only trend is that there is no trend. The Eurovision song is as highly specialised a creature as Miss World and, despite commanding our admiration for its wonderful laboratory freshness, the success of 'Save Your Kisses for Me' indicates little other than the baseness of public taste.

A mere fistful of years ago, I believed, and believed passionately, that the folks out there in Radioland would be tearful in their gratitude if they were to be offered the chance to hear, from time to time, rather more adventurous music than that which they are conditioned to expect to hear on Radio 1 – and, indeed, from the new commercial stations. With real regret, I have come to realise that this is not so. Some, even many, will be pleased, of course, but the hard truth of the matter is that the guys and gals want fishfingers, Tony Blackburn and television serials about, if possible, West Indian lavatory attendants whose wives have monstrous breasts. The admirable Beachcomber summed up the whole melancholy situation when he observed, announcing his retirement, that he could not compete in a society where it is regarded as the apogee of wit for a personality to introduce himself to his audience with a 'Good bloody evening'.

But let us get back to our waiting 450,000,000. What records are they buying? A quick glance through the international charts in *Cashbox* magazine is singularly unrevealing. In the Argentine, they are buying 'Para Piel de Manzana' by Joan Manuel Serrat, and 'Desearia Que Estuvieras Aqui' by the Pink Floyd. Italians are untrousering their embattled lire for 'Mina Canta Lucio', and 'Wish You Were Here' by the self-same Pink Floyd. The wily Japanese, on the other hand, are keen on Masato Shimon, Hiromi Oota and others. Among the others, in passing, is Hiromi Goo. I'd love to hear Hiromi Goo, wouldn't you? Our Eurobrethren, the French, care considerably for Genesis, Bob Dylan and Serge Lama. Actually, I considered the French song for Europe, given the standards established over the years for frothiness and an unreasonable gaiety, the best of a fairly horrific bunch. The evidence from *Cashbox* is that there are few trends of a global nature to be divined.

The truth is that, in 1976, the record-buying audience, considered as a mass (which is a fairly ludicrous way of considering it, I admit) is playing safe. Elton John has slipped slightly from his recent pre-eminence, but still shifts stacks of records. Folks can hardly contain themselves at the prospect of obtaining further 'product' (the record industry's word, not mine) from those two leading cake-mixers, John Denver and the Carpenters.

The market until recently designated as 'progressive' continues to distinguish itself by the extreme lack of progress in the development of its tastes. Although I have a lot of time for them myself – after all, I grew up with them – I am astounded that such ageing combos as The Who, the Rolling Stones, Led Zeppelin and the Pink Floyd still dominate the romping and stomping field. And, if you need further proof of the conservatism of current taste, then consider the appearance in our charts of several re-released Beatles' singles. Glad to hear them again, of course, but . . .

Around about now, I should be coming through with some sort of summary of the state of play, some glittering aphorism chosen for its elegance rather than for its accuracy. The trouble is that there are no conclusions to be made. There is, to be sure, a wealth of good music being made and recorded, but it lies, in the main, unacknowledged by playlist committees, unbought by the customer, and doomed, at best, to a half-life in some minority culture or other. Let us just say, for the sake of neatness, that, in these unpredictable times, the man on the street is settling for predictability. And who can blame him for that?

# Everything's Up to Date in Kansas City

*Radio Times*, 9–15 February 2002

---

IF YOU PASSED ME in the street you'd be unlikely to notice me, unless you thought, as so many apparently do, 'Isn't that Radio 2's personable Bob Harris?' If you had ever read any of the boys' adventure stories from the twenties involving lads named Jack and Spanish men o'war, though, you might, I suppose, notice my rolling gait. This is the result, alas, not of years spent hoisting top fo'gallants in the Roaring Forties but of a recurring corn on the bottom of my left foot. Over the years, foot specialists have offered to rid me of this, but my attitude is that the corn and I – and no, I don't have a whimsical name for it – have faced life's vicissitudes together undaunted and that, as it doesn't try to get rid of me, I won't try to get rid of it.

Nonetheless, every once in a while I go into Stowmarket to have the corn shaved down to a workable size so that I don't, for a while at least, have to walk on the side of my foot.

Last Friday a film crew was due at the house to harvest my thoughts on diabetes and, anxious to look at my absolute loveliest, I went to the chiropodist to have the corn trimmed. It was unlikely, I knew, that the film-makers would wish to film the bottom of my foot but, as I always say, TV people are capable of anything and you can never be too careful. Sheila had charged me with buying a loaf of bread before I came home, so I left the chiropodist and headed for the bakery on the other side of the inner relief road. This is a real bakery, baking real bread, and as such something of a local treasure. 'We'll have loaves in five minutes,' the assistant told me when I asked for a large white, so I went for a stroll across the river to the station, walking upright, feeling good, sensing that maybe spring was in the air.

There's a shop on Stowmarket station which years ago was run by a man who would, at the first hint that passengers were arriving to catch a train, lock the glass door, put up a hand-written sign that the shop was closed for five minutes and sit just inside the door, plainly visible and smoking, until the train had departed and passengers alighting had gone to the car park or into town. Nowadays things are very different, although memories of the old regime were strong enough that when the woman behind the counter wished me good morning, I was sufficiently taken aback to ask her to repeat what she'd said.

I left Stowmarket station to collect my large white in an impressively good mood and arrived home celebrating the essential goodness of humankind, the primacy of freshly baked bread, the joy of being able to walk properly, the admittedly rather watery sunshine and the miracle of Hubert Sumlin's guitar playing on the Howlin' Wolf CD I'd been listening to in the car.

I was singing when I walked into the kitchen.

'Everything's up to date in Kansas City,' I bellowed. 'They've gone about as far as they can go.'

Sheila looked alarmed. 'You OK?' she wanted to know.

'Just getting off on life,' I smiled.

She looked disbelieving.

# Extreme Noise Terror

*Observer*, 5 June 1988

---

'GET HIP TO VIC' is the strange device on the banners being carried by the ensigns of the Venue for Ipswich Campaign, the organisers of which move on to point out that apart from the odd performance in the upper room of an inn and the occasional concert at the Odeon, little happens to enrapture Ipswich's music-lovers. With the town experiencing something of a boom both commercially and socially and fired by the successes of similar campaigns in Norwich and Cambridge, Vic feels that, as it were, now is the hour.

This view was endorsed by a cook from a nearby American air-base who attended a fund-raising concert at the Caribbean Centre last week. 'Ipswich is a happening town,' he ventured. Living but a dozen miles due north of Ipswich myself, I know the town well enough to understand the significance of this opinion. Five years ago anyone placing the word 'happening' in the same sentence as 'Ipswich' would have excited the attentions of personnel in white coats. The concert featured a handful of local bands and was headlined by Extreme Noise Terror. Music between the acts was provided by Gibberish G on the wheels of steel.

The audience of several hundreds reacted with respect rather than solid enthusiasm to both the hip hop Gibberish played and to Johnny Benji, who chatted in a dance-hall style without resorting to too many of the banalities that make much contemporary reggae sadly uninvolving.

Perfect Daze, whose second release for Vinyl Solution is now available, turned in an enthusiastic if untidy set, after which we started to prepare ourselves physically and spiritually for Extreme Noise Terror. Now, I speak as a man who has played records by Extreme Noise Terror on the radio and featured the band in session, but nothing had prepared me for the impact of the band live. Sheila, my wife, who has seen a band or two, stood open-mouthed as they raged. Gibberish G, rooted behind his turntables, shook his head in disbelief.

Propelled by the drumming of Mick Harris and fronted by the lead singers Dean Jones, who delivers portions of song in an unearthly growl, punctuating his work with arm movements which give the impression he is lobbing grenades into the crowd, and Phil Vane, who produces from behind a veil of hair a hoarse bellow which surely has its origins deep in the mists of prehistory, Extreme Noise Terror play so fast and with such fury that I found myself turning to Revelation for suitable comparisons.

'And there were voices and thunders, and lightning; and there was a great earthquake, such as was not since men were upon the earth, so mighty an earthquake, and so great,' seems to go some of the way to capturing ENT in performance, and only my keen interest in their work stopped me from stepping into the street outside the Caribbean Centre to see whether the sun had indeed become black as sackcloth of hair, and the moon become as blood.

Ipswich is the place to be.

# Fab Pic Contest!

*Sounds*, 22 October 1977

---

BUT FIRST, this colerm, beloved of millions the world over, announces a THRILLING COMPERTISHUN open to BOYS AND GIRLS OF ALL AGES. Study the series of FUN SNAPS of myself and flatulent Radio 1 producer John 'Petals' Walters, then consider how tricky it must be for one ageing fatty – let's call him A – to replace a second ageing fatty – let's call him A as well – in the automatic snapshot booth for the third frame – and then CHANGE BACK AGAIN for the fourth. I intend to offer, FROM MY OWN POCKET, a prize of £25 (YES, that's twenty-five pounds) to the group of individuals who can get the largest number of DIFFERENT folk into one set of booth photographs. Sounds like fun, doesn't it?

Some contestants may also have the questionable pleasure of seeing their efforts reproduced in this paper. (Actually, I haven't checked this with the Editor yet, but I do have a number of indiscreet letters he wrote to Olivia Newton-John and anticipate little trouble from that quarter.) So go to it now. Incidentally, only humans are eligible, so there is little point in carrying in your collection of moths and hoping to win the money that way. Petals and I used the booth in Soho Market, in the middle of London's bustling West End, and alarmed a substantial number of Belgian tourists with our infantile goings-on.

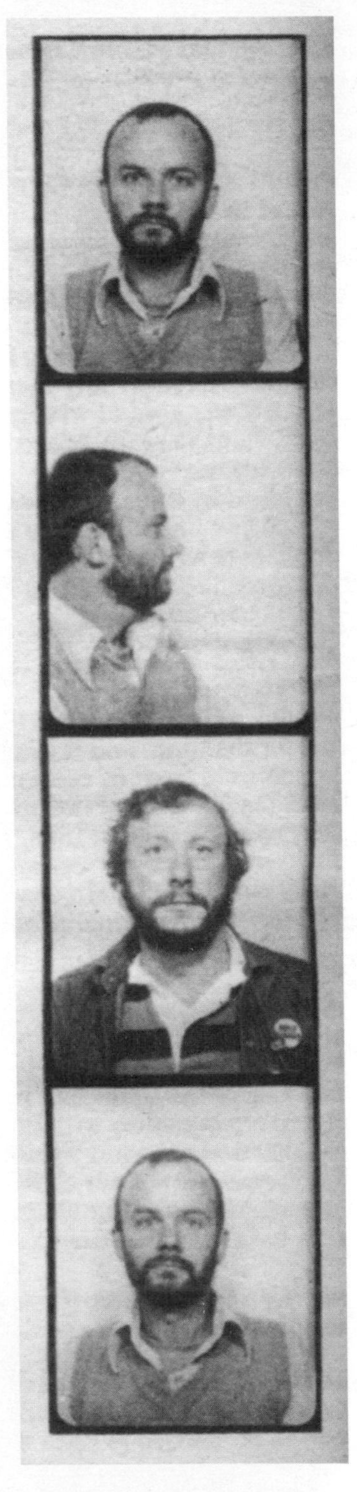

But seriously.

Lupescu, the blood now dry on his domed forehead, leaned forward and peered through the rain-flecked windscreen of the speeding Avocado. Bidet should have reached Vienna by now and handed the all-important briefcase to Vesper. As he mused, Lupescu turned around slightly in his seat so that he could see Jill, still bound and gagged in the back seat. He smiled to himself with satisfaction. Vesper would be well pleased with his work this night – might even allow him the pleasure of eliminating Cowley once and for all. His smile broadened at the thought.

The lights of a small town drew near in the night. It was quarter to two in the morning. Lupescu checked his watch. It would be seventeen minutes past five in Seoul, exactly half past eleven in Buffalo, and a few seconds short of twenty-two minutes to four in the afternoon in Hyderabad. The careering Avocado was well ahead of schedule. Road signs told him it was eight miles to Dumfries.

In the back of the car Jill struggled soundlessly with the tape that bound her wrists. Unless she could free herself before Lupescu made that fatal right turn into Prestonpans Crescent and drew up outside number nineteen, Vita and Merk would have died in vain. She felt the tears welling in her eyes again as she thought of poor Nod lying spreadeagled in the frozen mud of the Norwegian forest. Had he discovered, before he died, the truth about Yvette? No one would ever know.

The above paragraphs are included as a sop to the simply one or two regular readers who have persuaded their mums to sit right down and write to enquire of me 'What then has happened to *Of Buckskins and Buggery*?' Well, dear, dear friends, the final gut-wrenching, soul-searching manuscript now rests in the palsied hands of my publishers and I am hopeful that I will have wonderful news for you in a very few weeks' time.

Kid Jensen has just sauntered into the office, and he came straight over to me and asked me, rather tersely I thought, why I hadn't changed the roller towel in the toilet at the end of the corridor. Once we had straightened out this silly little misunderstanding – we're awfully good friends really – he begged me to make enquiries about the following small ad which appeared – I think that's the right word – in last week's ish . . . 'Kid Jensen 25th tape wanted. Alan Jenkins, 2a Llynll Street, Bridgend, Glamorgan'. What Kid wants to know – and I'm with him 110 per cent of the way on this one – is this: what is a '25th tape'? Answer me that if you're so damn clever.

# Fab Pic Contest 2

## John Peel Gets the Bird

*Sounds*, 5 November 1977

---

THERE ARE TIMES, are there not, when, quite frankly, one wonders? I am wondering even as I write these words. And why am I so wondering? Well, since you ask me I shall tell you. You, dear reader o'mine, may recollect that a week or so ago I reported in these pages on a gig the Peel Roadshow (motto 'Wholesome Entertainment for the Entire Family Unit') had done in Walsall, at the West Midlands College of Education. In the course of my remarks I observed, if you will permit me to remind you, that 'there were those present who were both affable and generous, but the prevailing impression was of doing a disco for the West Midlands Home for the Terminally Morose'. Hardly hurtful, I reckoned, and I further reasoned that the dowdy folk who objected to my saucy raillery and fun would hardly be readers of *Sounds*.

But I was wrong.

Imagine my surprise when, a week later, I had a call from the local Walsall paper asking for my comments on reports that some of the victims of my elephantine barbs had been sufficiently wounded by them to form themselves into an action committee, and thus bring the whole matter to the attention of the press. To the man from the Walsall paper I observed that the complainants must be even more dour and humourless than I had hitherto imagined. The kind of horrors who start sentences with 'I consider that I have a good sense of humour, but . . .' When I put the phone down I felt that here was an end to the whole preposterous business.

Again, I grieve to tell you, I was seriously in error.

Some of the boys and girls of the West Midlands College were, it transpires, *so* distressed by what the naughty, nasty man had said to them and about them that they scampered home after school and told their mummies and daddies all about it. The enraged Mums and Dads of Walsall, keen to protect their little ones from such wickedness, have organised sufficiently to have written to their MP, the man who stepped into the boots abandoned by John Stonehouse.

Now it so happens that this man is the only Member of Parliament, as far as I know, with whom I was at school, and he has written to me a letter, on House of Commons notepaper, asking, with some embarrassment, for an explanation. It had been reported to him that I 'fairly laid into the students, describing them as "morons" '. Oddly enough 'morons' – because the word has a real and very exact meaning, and I don't employ it lightly – is possibly the only word in the entire vocabulary of popular abuse that I would not have used. Arseholes, for example, yes; morons, no. Perhaps the distressed gentlefolk and their parents misread the word 'morose'. Who knows? Who cares?

Oddly enough I am now, in the murky light of today's developments, awaiting a letter from the member for Wolverhampton – or whichever part of that city includes the home of a young man I met recently in Norwich. This latter worthy attended a recent Roadshow gig in Norwich, and was good enough to take time to hurl a dead pheasant into my testicles as I addressed the assembled company of youthful jivers. Two adjacent chaps set off in pursuit, with your old chum wheezing along behind them.

When I caught up with the action the bird-handler had been wrested to the ground, and I proceeded to question him closely on his motives for so assaulting me. He could only attribute his actions to the fact that he was drunk, and his friends, who were later to hang about the exit from the place where we were working with the intention of doing me harm, seemed to think this was a wholly satisfactory reason. They even accused me of over-reacting to the attack, and by the end of the evening an 'eye-witness' had been found who was prepared to state that I had put the boot in on the pigeon-fancier.

As I was wearing gym-shoes this would have been remarkably ineffective anyway, but so rapidly do such stories take on new and hideous facets that I expect by now it is generally accepted that I attacked the man, who has been kind enough to offer me a game pie, with a regiment of foot and air cover. In vain I tried to reason with the Wolverhampton man's comrades, asking whether they thought that, say, the average citizen out for an evening beer would, if struck in the rude bits with a dead fowl by a complete stranger, take it as a mere passing pleasantry (*pheasantry? – Ed.*) and ignore the whole business. They gave scant attention to my sophistries, preferring, as I hinted above, to plan further assaults on my person, presumably with the hope of persuading me to receive carrion in the groin with a grateful smile in future.

Do these things happen to Noel Edmonds? Or Paul Burnette? And what can I expect from the three remaining gigs in my pre-Yule British tour? As a man who hasn't been in a fight since the age of seventeen (I won that one though) I live, I assure you, in constant fear and trembling. I shall be pleased when William (A Whirlwind) and the Impending Edward James Dalglish Peel/Ravenscroft are large and vicious enough to protect their poor Dad.

To other and jollier matters. Response to the vital emotion-charged competition announced here a week or so ago has been,

I think it is fair to say, modest. Someone *will* keep sending me extremely rude pictures of key parts of the female anatomy, and the only serious entry has come from Brentwood, Essex. I drive through Brentwood several times a week and never imagined there were people in that grim dormitory capable of such tomfoolery. However, sixth formers of Shenfield School clearly are, and printed herein is their entry. Unless someone comes along with more than fifteen different individuals portrayed in a set of four machine photos, they will win the £25 in seriously soiled banknotes. I'll give you until the end of November to do it.

Off I go – and keep that confounded ptarmigan to yourself, you twat.

# Fab Pic Contest 3

*Sounds*, 31 December 1977

---

FRIDAY 9TH DECEMBER. 5:55 a.m. The telephone rings. Peel, his strong, clean, honest, British face cupped in his hand, lies prettily aslumber in London W2.

The phone continues to ring until our hero, waking, springs panther-like from the bed and strides across the room to answer it. A female voice greets him, tired, distant. 'Can you come home, you paragon of virtue? I think it is happening at last.' Pausing only, in the public interest, to fling on some clothes, Peel hurtles out of the building and into the driving seat of his van.

An hour later he is observing the speed limit along the mighty A12, his ever-vigilant eyes scanning the roadway ahead, the slight breeze from the open window tugging at his tousled hair. My god, he looks good enough to eat.

Another hour (that makes two (2), students) has passed before he skids to a halt at the end of the sweeping driveway that leads to the ancestral home of all the Peels. Rushing into the house, brushing aside dogs, cats and staff, he bursts into the with-drawing room, to find his wife, the Lady Pig, moaning softly on the Louis Quinze sofa. The contractions, she gasps, are but ninety seconds apart.

Peel turns pale beneath his tan and directs Gabs, The Appalling Red-Haired Sister-in-Law, to help said Pig into the van outside. This done he turns his wheel for Ipswich. As he reaches the outskirts of this fairest of cities, he runs afoul of rush-hour traffic. Being, as we all know, a man of action, he turns on the headlights of the van, and takes to the pavement, sweeping past a three-hundred-yard-long line of traffic delayed by road works. How wonderful it is for us to be able to see such a decisive chap at work, eh?

At 8:55 he draws up outside Heath Road Hospital and assists the Pig, now in considerable pain, into the building. He hardly notices the admiring glances of the trainee nurses as he manfully summons the lift.

Forty minutes later he stands in the delivery room, clutching to his chest his spanking new daughter, the girl who will one day be the toast of all England. The expected Impending Edward, it seems, is further biding his time, waiting perhaps until Kenny Dalglish rediscovers his goal-scoring form.

Since the events described above we have been unable to find a suitable name for this most elegant of children, despite debating the issue far into the night in rooms lit with but a single guttering candle. William (A Future Prime-Minister) calls her 'Bady' (do not, dear proof-reader, correct that to 'baby'. Thank you) and the Pig and I call her, for the time being, Childe B. I have already warned her against encouraging the unwelcome attentions of men such as myself. Forewarned is, as I always say, forearmed.

This week I have made my way into the *Sounds* office, to see the boys and girls and to shower them, as is my custom at this season, with pieces of eight. I can scarce tell you how horrid it is to see them all rolling on the floor fighting for the coins, kicking, biting, spitting and scratching in the frenzy of their greed. While the noisome crew struggled I collected a vast mound of correspondence sent in by you, my adoring fans, and made good my escape. Let's see what you have to say, if I can decipher your dreadful handwriting.

Firstly, there are several more entries in our futile How Many Wastrels, Degenerates and Drunks Can You Cram Into A Photographic Booth? competition. Although the contest was officially closed weeks ago, many of these entries were posted before the closing date and must therefore be considered valid.

Here is a note from a Jo Betts of Leicester. She (or he) sends but two pics, rather than four, and these have a measly four souls (and a stray hand) on them. Hopeless, I'm afraid. Recycle. From Norwich write the fifth form of the Hewett Comprehensive School, represented by a Daniel O'Donoghue. He, his sixteen friends and an unclaimed ear, were chased from the bus station where the snaps were taken by an army of constables, and Daniel feels that this humiliation entitles them to at least a share of the prize money. Wrong, Danny baby. Plucky try though. If the nutters in the pictures are your friends, Dan, I think you're associating with the wrong sort of people. Trust me.

And so to Cramlington, Northumberland. (*Where? – Ed.*) Someone pretending to be called Neil McAnany sends a strip of four boothographs depicting twenty-one dreadful looking humanoids. They are all male. What sort of lives do you lead up there in Cramlington, I'd like to know? But, like it or not, these twenty-one move into the lead.

Nick Morgan of Echington, Nr Pershore, has attempted to deceive the judges by mailing in four separate photos of a total of twenty-three of the indolent pupils of Prince Henry's High School in Evesham. Nice try, Nicky, but who's to say that these four pictures are not the best of dozens of different strips? Hold on: I've just rechecked the pictures and the same people appear in each of them.

But all of these puny efforts pale before that of members of Churchill Hall at Bristol University. They have demonstrated that everything your long-suffering parents say about students is absolutely true by cramming twenty-three different persons into four unseparated photographs. Either they fixed the machine or they are all half-crazed with drugs, but, either way, they have managed to capture the prize. Hats off to the swine.

P.S. Childe B is now Alexandra.

# Faces

## A Curse on Reporter Erskine . . .

*Disc*, 21 April 1973

SINCE LAST WEEK complete strangers have been stopping me in the street and asking, 'Who, then, friend, is this Peter Erskine person?' Regular readers (and there are over 250,000,000) will know that the man Erskine is the Herbert who reviewed the epic *Ooh La La* last week and was less than kind about it.

Several of the folk, who stopped me on the busy highroads and by-roads of our country, were so distraught that they were anxious to find this Erskine and strike him many times with the jawbone of an ass and other such devices. So far I have managed to restrain them.

I may relax my vigilance before the week is out. In the Gothic splendour of Peel Acres we have been working on a device into which we can introduce Erskine which will force jets of Fairy Liquid up his nostrils while pebble-dashing him and bricking up his other body orifices.

Several villagers have been spoiled in the course of our experiments but what is the loss of a few citizens compared to the benefit of mankind that will be the inevitable consequence of our turning Erskine into a lamp-standard on the Hendon Way?

Readers who feel, as I do, that the Faces' *Ooh La La* is as dandy a wee record as we're likely to get this year, will be gratified to know that Peter Erskine is a hairless midget with so many facial tics that his face moves ceaselessly and silently like a pool of boiling mud.

His opinion is without value. Compared with him Rosalind Russell is a pillar of society, a human being of virtue and charm – and one can say no more than that.

This weekend (and I had hoped to bring you a detailed report but you'll have to wait until next week, so order your copy now for the demand will be enormous) I will be pointing the bows of the faithful Land Rover Friday up the M1, splicing the top fo'-gallants and flying before the breeze to Sunderland to see the aforementioned Faces.

I've always enjoyed Sunderland, but Sunderland with the Faces is a special treat (their last gig there was the best Faces gig I can remember), and Sunderland with the Faces a week after their well-deserved stomping of Arsenal could be something to change the ebb and flow of history and stay the stars in their courses.

From Sunderland, having battened down the hatches and hoisted the royals, I head for fame and fortune in Scotland. I have new 8-tracks of such sensual delights as the Floyd's *Dark Side Of the Moon*, *ELO2* and Roxy Music's *For Your Pleasure*, so the hours spent patiently following Highland cattle down cart-tracks should pass quite easily. I don't have an 8-track of *Ooh La La*, though.

Perhaps some record company person will take time off between banqueting and travelling to and from London Airport with obscure guitarists to send me one. I'm a demanding little bitch sometimes.

# The Fall

*Observer*, 24 August 1986

---

OOK NIET BIJ *John Peel . . . hij schijnt onze muziek niet te mogen.* My Dutch is a bit rusty, but this, as I understand it, is a summary of Claw Boys Claw's attitude to my lack of enthusiasm for their records. Having seen the band play at last weekend's Waterpop free festival at Wateringen, a kilometre or so from downtown Delft, I have little hesitation in extending this lack of enthusiasm to embrace their live performance.

Claw Boys Claw enjoy considerable support in their native Holland and won some rather muted praise at Waterpop from a contingent of Englanders, the spawn of Eurofolk based in The Hague, but all they have to offer is a variety of pub rock. A small voice within that will not be stilled insists that little would please them more than to be described in print as 'the bad boys of Dutch rock 'n' roll'.

The festival site, set amid trees and fringed by a narrow stream on which ducks have no hesitation in sporting, is almost abrasively picturesque. It played host on Saturday to several thousand music-lovers, most of them dressed in black, a sprinkling of curious locals of all ages, and several increasingly desperate lost dogs. The event is sponsored by the local business community, in combination with the Wateringen council, in a demonstration of concern for the appetites of their young rarely, if ever, matched in Britain.

Following Claw Boys Claw, we stretched out on the grass to await the arrival of The Fall. To be frank and manly with you, I have been rather dreading writing about The Fall, my favourite band since The Undertones ceased trading, because my enduring admiration for their unwholesome racket is not something that can sensibly be analysed.

For a decade The Fall, fronted by the hunched and unsmiling Mark E. Smith, a man with the least enviable wardrobe in all of Northern Europe, have kept their music clear of mere fashion, untainted by opportunism. At Wateringen their gratifying refusal to communicate with us on anything other than a musical level – Mark E. Smith's only audible spoken words were 'Let's do "Bournemouth" and get out of this place' – resulted in a more direct communication than any oafish attempt to get us to sing along or clap our hands, devices which have always seemed to me to serve to increase the distance between performer and consumer.

The Fall's apparent lack of interest in us somehow drew us in as co-conspirators, and, as the catalogue post-punks in Cure and Cult T-shirts inched away looking resentful during one of the band's longer and more mournful numbers, I was moved to chortling out loud. The Fall's set was drawn in the main from their forthcoming LP, with 'Terry Waite Says', 'Pharmacist', 'Bournemouth Runners' and 'US 80s 90s' being particularly striking.

They also performed both sides of their current single, a record which has manifestly failed to leap into the charts – I yearn for the day when I can introduce The Fall on *Top of the Pops* – and as an unexpected encore, as we made our way from Waterpop 'neath the lengthening shadow of a vast windmill, last season's teen-terrific 'Cruiser's Creek'.

This most singular group remains, in the words of the festival's free newspaper, 'steeds weer iets onders'. Couldn't have put it better myself.

# Bryan Ferry

## John Peel Catches the Ferry to Oblivion

*Observer*, 22 January 1989

---

A FLYER DISTRIBUTED on behalf of the lager manufacturers promoting Bryan Ferry's current tour alerted merrymakers attending his Wembley Arena concert on Thursday to the fact that they were about to witness 'the ultimate performance' by 'Britain's premier rock artist'.

I first saw Bryan Ferry with Roxy Music when they, together with Genesis, played the Hobbits' Garden in Wimbledon. Afterwards I gave the band fatherly advice about careers in showbiz. Bryan's appearance at Wembley demonstrates how little attention he paid to this advice. Following a distinctly rum introduction tape – lowing, muttering and clanking apparently played at the wrong speed and backwards – Bryan arrived somewhat diffidently before us. He wore red trousers which, from my position halfway down the hall, appeared to have the exciting, new, low gusset. Either this, or Bryan's a funny shape.

Much has been written about Ferry's inability to move authoritatively on a stage, and it is true that he dances as Prince Charles might when surrounded by capering persons in penis gourds during some half-baked tourist ritual on an atoll protectorate. But we love Charles for his clumsiness, and we love Bryan for his, too.

I am, alas, not as familiar with the Ferry catalogue as I should be, but I was hoping for a few Roxy Music faves or a torch ballad or two that would give Bryan the chance to smoulder at us. Instead we got a barrage of pop dance pieces, many of which seemed just to peter out rather than reach a conclusion. There was one, for example, which had our boy – and the trio of female singers behind – going 'bum bum' until I feared for my reason. Then there was another where he went 'a ha ha a ha' for a comparable spell. Mind you, he did whistle as well. There should be more whistling, in my view. For anyone planning to be the rock Ronnie Ronalde the time is *now*.

Noble and not unsuccessful attempts were made to make the songs sound a touch more riveting by avoiding the more obvious accompaniments. Thus one was ornamented with the sort of string sound you get in Indian film music, another with an electronic approximation of the myriad magical sounds of the rainforest. A third featured the type of noises that appear on mood music LPs under the title 'Spring In Rio'. A low point was reached with an oafish reading of 'Jealous Guy', a song which in John Lennon's original is built on a tangle of anger and regret and here had neither.

Mass observation techniques employed on those about me, many of whom arrived up to forty minutes late and smelt, barely credibly, of baby powder – is there some alternative use for this of which I am unaware? – revealed that toes remained untapped. The young marrieds alongside me continued their dinner conversation during all but the loudest passages and eventually, bored with them and, I fear, with Bryan, I left.

# Football

## John Peel Takes to the Football Field Again!

*Disc*, 7 December 1972

ER. AHEM. Well, hello. That is to say, I mean . . . I'd better start again.

This year, you will have noticed, I have spoken very little about my prowess and numerous successes on the football field. I will tell you why. The only team I played for last year was our very own and dearly beloved Radio 1 Stanley.

The problem is that Radio 1 Stanley play at the crack of dawn on a Sunday morning and some one hundred plus miles from our fortress home. So earlier this year I applied to join our local village team and even filled out a series of amusing forms in order to become a proper and legal footballing person.

There was never any prospect of my turning out for the first team but I had high hopes of a regular game with the Reserves. You can imagine the excitement here at multi-million-dollar Peel Acres when I was selected to play for the Reserves several weeks ago.

I will not pretend that I was a sensational success but, I felt certain, the all-powerful selection committee would take into account the fact that I was playing my first game in eighteen months and would give me another chance.

Alas, it was not to be so. Every once in a while I get a form through the mail telling me that I am chosen as a reserve for the Reserves and so, with Vaselined feet and bandaged legs, I go each Sunday morning in the hope that three or four people will be stricken with some mysterious, incapacitating but temporary disease and I will be able to be substitute.

It has not happened yet and John – my friend in the team and the man who seems to score all of the goals despite his curious physique – he makes me look almost girlish by comparison – tells me that it is not likely to happen.

Perhaps I should go and ask for a transfer and see if there are any teams of enthusiastic but inefficient people who would give me a game who don't live hundreds of miles away. I watch *Match of the Day* regularly, except when Arsenal are featured (which seems to be on the weekends when Leeds or Manchester United are not featured), and I know how it should be done. I've got my own shorts too – and socks.

Still they come, from Poland and from Peru – the letters from eager young girls who want a Full Colour Free Poster of my good self. But our Editor, a stubborn and wayward fool of, I might as well tell you, virtually no formal education, and a face so pock-marked that small children run and hide when he stumbles drunkenly by, will not permit it. Instead we are fed a dreary succession of ne'er-do-wells and degenerates. I feel the time is coming for a workers' take-over in the *Disc* office.

In between the letters urging us to come on and put Scott Walker back in the charts where he belongs and letters explaining with a certain amount of venom that Bread are better than the Pink Fairies and vice versa, there was a brief note from a Jim Smith, of Wood Green, thanking me for bringing the praiseworthy Home to his attention.

Now this, frnakly, is the motive behind everything I do. Well, if you insist, not so much frnakly as frankly. Everything that is except this column. When I write about music, people write and say 'don't write about music' and when I don't write about music they write 'why doesn't John Peel write about music.' If you don't make your minds up soon I shall have to take my rhinoceros hide whip to you.

As I was saying, the programmes with which I'm involved, and to a certain extent the reviews in this journal, are aimed at turning y'all on to some musicks that you might not otherwise investigate.

So I'm glad that I helped to bring Jim Smith (obviously an assumed name) and Home together. If you'd listen to the programme instead of watching that damn-fool TV I might be able to do the same for you.

Last week I gave a hitchhiking gent a lift up a bit of the A1(M). After a while the conversation turned to rock-a-boogie and he said 'Hey, aren't you John Peel?' I smiled modestly and said 'Well, yes.' 'Are you still on the radio?' he asked.

You'll find him in a shallow grave on the roadside near Baldock.

# Foxes

*Radio Times*, 19–25 October 1996

---

HOW ARE YOU at killing things? I'm not terribly good at it myself. Years ago and just down the road, I came upon a duckling that had been irreparably damaged by a cat and decided to put the stricken creature out of its misery. I still remember in my dreams how my attempts to chivvy the bird into the valley of the shadow of death with the sharp edge of a shovel met with heart-breaking resistance. Worse still was the Labrador I hit one night about twenty years ago as I drove home on the A12. The dog appeared from nowhere and all I saw of it before I hit it at seventy miles an hour was it walking towards the hurtling car wagging its tail. If I close my eyes, I can still see that Labrador.

Years later, when the children were eight, six, four and two, we often stood at the kitchen door to watch a fox cross the field, outlined against the sky. We found something wild and beautiful in that. We were horrified one afternoon when the local hunt appeared in the garden and killed the fox in front of us. A phone call that evening to the hunt secretary, a man we knew and liked, brought forth the suggestion that if we had not wanted the hunt in our garden, we should have notified him first. I asked whether this intriguing concept could be extended to, say, my burgling his home if he failed to ask me not to do so. It turned out that it could not. What then, I wondered, about my exercising my dogs in his garden? This also was unacceptable.

I thought it unwise to explain to the hunt secretary (he was also our family doctor) that my sympathy for the fox stemmed, in part, from my childhood, when my brother Alan – small, smart and infuriatingly cute – had this recurring nightmare that would wake him night after night. It seems he was being pursued by a fox in a well-cut red coat. The fox, Alan explained between sobs, gave every indication that it believed my brother would make a delicious and nutritious snack. I loved that fox.

Also, like many other people, I have never really rid my head of the idea that foxes return home to snug dens fitted with reproduction pine furniture and matching gingham curtains and tablecloths, in which plump vixens in bright pinafores prepare tea, and boisterous cubs want to be read stories about sea captains.

For these reasons, I approached *Twenty-first Century Fox* with some unease. Also, I am aware that few other things are as likely to trigger overheated mail. One side will send me badly printed but nonetheless harrowing pamphlets illustrated with photographs of grinning rustics attacking saboteurs with staves and cudgels; the other will write letters telling me that hunting is the best, the only way to limit the damage caused by foxes – from this film, it seems it isn't – and that the centuries-old hunting tradition and spirit were the very things that helped us so convincingly to win the Falklands War.

As Dr David MacDonald of Oxford University points out, society is very, very inconsistent in its ethical judgements, and doubtless many of those opposed to fox hunting keep a cat, turning a blind eye to the beastly things it will be doing to dear little robins and mice. I myself drive a car and choose to ignore the fact that as I speed about in it, I leave in my wake a nightmarish trail of dead and dying creatures, mainly insects.

In *Twenty-first Century Fox*, Julian Pettifer draws on his wellingtons and meets farmers and hunting people, nature-reserve wardens and gamekeepers, but at the programme's end, this liberal softy was more confused than ever. Is there a humane way of controlling any animal population? ('No,' suggests one of the programme's participants.) Is it right to interfere with natural processes and protect one species at the expense of another? Can it really be true that gamekeepers kill over 100,000 foxes a year to ensure a plentiful supply of semi-tame birds for people to shoot? Whose side are you on when you learn that 2 to 3 per cent of free-range piglets, born to be devoured by you, are taken by foxes? The piglets'? The foxes'? The farmers'? *Twenty-first Century Fox* has no hesitation in leaving these questions unanswered. But then, they are unanswerable. Not that that is going to stop people writing in with answers, I fear.

# Glastonbury

*Guardian*, 2 July 1993

WHEN MY SON William was four or five his favourite sport was to kick a football as hard as his little legs would allow, past his groping father and into the ditch. This game was called, as you might imagine, Ball-In-Ditch.

At last weekend's Glastonbury Festival I was forced to play a rather horrid variant on Ball-In-Ditch when I retrieved a football loaned by William's brother, Thomas, from a brook into which hundreds of music-lovers upstream had urinated. I consoled myself with the thought that this beastly experience would enable me to speak with unique authority on festival sanitation.

In addition to these journalistic duties, I was at Glastonbury for a live programme on Radio 1 FM and to act as occasional compère on the NME stage. I have to tell you that in 1993 the role of compère is not an easy one. Carefully chosen records are faded to a whisper by sound engineers checking microphones and really rather clever introductions go unheard as band after band opts for taped alternatives.

Despite a 7 a.m. start from Stowmarket, our little party of merrymakers arrived on site too late to experience Rolf Harris, which I deeply regret, but in time to hear the Mexican band Maldita Vecindad. I was able to combine two of my roles in one in listening to them from the backstage Selwood Two Seater, which, like the band, was clean and functional. Maldita Vecindad are well enough drilled in rock's clichés to keep them in support roles for years. The Selwood's cubicles are perhaps a little snug.

The likeable Mexicans were followed by the Auteurs and the North Carolina moptops, Superchunk, the latter turning in a snappy and unselfconscious set in sharp contrast to much of what was to come. In 1966 I failed to see the Velvet Underground in Hollywood, turning aside at the last moment and going to hear the Seeds instead. In 1993 I failed again to see the Velvet Underground, this time at Glastonbury, but only because I was too far from the stage and today's young people are too tall. Before growing impatient, I heard three numbers which sounded as if they were being performed by a highly competent covers band, probably called something terrifically clever such as Velcro Underground. Later I fulfilled a long-held ambition by kicking footballs from the stage for – and, quite frankly, further than – Teenage Fanclub. Later still I should have watched Suede, especially since it was rumoured that David Bowie would sing with them. (He didn't.) Instead I wandered about the huge site, the last rays of the sun dancing on my hair, sampling food and drink – including a hellish product identified as damson wine – and inspecting goods offered at stalls bearing such names as Armalite Sparrow, the Invisible Clothes Shop, the Silly Fish Shop and Stupormarket. With Andy Kershaw and our friend Tony, I sat on the top of a hill watching a distant tight-rope walker and a yet more distant fireworks display. In my heart of hearts I knew this was better than Suede.

On Saturday morning the Selwood was blocked and awash so I took my custom to a Site-A-Loo, a Tardis-like contraption, roomier than the Selwood but with certain aesthetic drawbacks I am reluctant to bring to your attention. First up on the NME stage was Les out of the Vic Reeves programmes. His astonishing performance included puppetry, Golden Keyboard Moments, a ladder and simulated farting. Next came Adorable, the

Rockingbirds, Eat, Dodgy and Verve, all competent but strangely uninvolving. Despite some hummable tunes, they seemed record company bands, ill at ease and overly theatrical. Desperate for effect, the singer with Eat showed us his penis, but somehow you had known he was going to do this. Grumbling, I climbed the hill to the acoustic tent with Andy and Tom, past ranks of jugglers, a youth who was trying to finance a trip to India through the sale of part-worn joss sticks, and the occasional knot of angry looking men with bruised faces, portable phones and, as often as not, Mancunian accents. What we found at the top was for me the musical highlight of the weekend, a set of unfettered charm from the Irish accordionist Sharon Shannon and her band. This, Andy and I agreed, was what Glastonbury was all about, rather than the self-conscious waving of penises. Afterwards Andy recorded a couple of tunes for the programme we were to do that night and even that small, private performance set onlookers to dancing.

The NME stage on Sunday seemed to be given over largely to bands clad in convoy chic, merrily a-prancing either barefoot or in boots that looked as though they had been prised from the feet of victims of the first battle for Ypres. Too many of the songs had revolution in their lyrics but not in their music. 'Have a knees-up. Get brown,' urged Back To The Planet. Perhaps I should have done. Instead, depressed, I walked to the Pyramid stage where Billy Bragg was deputising for a sidelined Nanci Griffiths. As I arrived, Billy was singing 'Mr Tambourine Man', a selection that, had he been busking, would have earned him a reproachful glance and no more. He finished a typically relaxed, funny set with a duet with Paul Brady on Woody Guthrie's 'Do-Re-Mi'.

Billy gave way to a short man with a face like a member of the post-war Labour Government, all unforgiving eyes and down-turned mouth, radiating solemn purpose. This was Van Morrison, and looking at his audience, I pointed out to Boy Kershaw that although we older pop fans may be ugly and have sore bottoms, our age has enabled us to hear some magical stuff. That night we were back on the hill, talking TT racing again as the sun was sinking down, as Donovan once put it, 'behind the tattered tree'. Then Donovan started singing right behind us and we moved hurriedly back to the plain to drink beer in the night mist with Mixmaster Morris. Morris is a high-priest of ambient music and he told how the musical future would be: anonymous, domestic, personal, star-free. I asked him what he thought of British guitar bands and he laughed. A hundred yards away the Selwood was still blocked.

# God Save the Queen

## What the Papers Say

*Sounds*, 18 June 1977

---

THIS WEEK'S KEENLY ARGUED discourse is written during the England vs. Argentina match. William (A Small Boy) and I are concerned to see that our Latin-American amigos have fielded a defender named Killer, who has just savaged one of our plucky British boys. Can it be, we wonder, that we are witnessing the birth of new wave football?

Later: Haven't got very far, have we? Five minutes to half-time and only one scrawny, undernourished paragraph to my credit. I was hoping that as the final whistle went I could tear this lot out of the typewriter with a great bellow of triumph, but at this moment in time (I got *that* from the television, you bet) it does not look promising for me.

William (A Small Boy), who is supposed to be suffering from German measles (the Boche's reprisal for the European Cup Final, I shouldn't wonder), having patted a liberal coating of asparagus soup about his corporation, has retired to bed. He explained before leaving that he was disgusted at having to watch Ray Clemence perform his usual miracles in the middle of what appears to be the overspill from a municipal dump.

But I promised you several weeks ago that I would make no

further mention of football in these columns. So let us move on to other and weightier matters.

Tommy Smith MBE! What an impertinence! (I fear I owe an apology to the more sensitive and doe-like among you for all these exclamation marks – and so early in the column too. When I was but a beardless youth, already, at thirteen, the toast of the warm-hearted women who sold oranges in the Strassestrasse, and studying your curious language in Vienna under dear Professor Zwanzig, I was taken to one side and warned that only braggarts and drunkards employ the exclamation mark in their writings. And this is so. But I have strayed from my theme.) In the paper only a fortnight ago I left pretty damned explicit instructions that Tommy Smith was to be knighted, and these instructions have been ignored. I am not at all pleased, at all, at all, I can tell you. Not pleased.

Mein Gott! What a night we had last night. We'd invited one or two of our very closest friends round for a drinky – not many, certainly no more than eighty or ninety – and we were dancing, drinking, copulating and discussing slipper-shaped infusorians – such as the paramecium – until dawn. Our neighbours complained to the local police on three separate occasions because they were unable to sleep for the popping of champagne corks, some of the world's most beautiful women took the opportunity to sample the myriad delights of my body, and we had to ask the Pink Floyd to play a late set for the dancers on the battlements. They weren't keen to do so, of course, having already played for seven hours, but they soon changed their tune when I sent Mrs O'Kelly around with another crate of light ale.

Well, if you must know, we stayed in and watched television. It was pretty terrible too. These days, what with William (A Small But Disconcertingly Powerful Boy) on the premises and Edward (An Embryo) on his way, we don't get around much any more – cute title for a song, eh? Trouble is, with radio programmes to do at nights, I don't get out much during the week either. Life, sometimes I think, is passing me by on the other side. Do you ever feel that?

This morning, having lightly breakfasted on a tangerine and half a swallow, deep-fried in its own fat (you really must remember to ask me for the recipe sometime – delish!), I turned what I have come to know as 'my attention' to the morning papers.

In the one I read of the end of the twin Dutch sieges, and of the marvellously expensive weekend we cringing taxpayers have decently laid on for the Commonwealth prime ministers and their wives, the former a group of bigots and crackpots of such rare accomplishment that they have managed, by a marvellous process of self-delusion, to arrive at the conclusion that the most vital issue before them is that of which countries should be permitted to go out and play during break, rather than what should be done about Idi Amin, contentedly butchering the boys and girls back home.

Then I turned to the *Sunday Mirror*. 'Punk Rock Jubilee Shocker' is the headline taking up most of the front page. 'Lawksamussy!' I think to myself, 'what can have happened?'

Well, a quick study of the piece crouching below this splendid headline reveals that the *Sunday Mirror* have discovered that a teen-beat group named the Sex Pistols have made a record called 'God Save The Queen'. Did any of you know that? Shocking, isn't it? It seems that these Sex Pistol johnnies have 'attacked' Her Majesty with their opening lyrics. The *Sunday Mirror*, bless its little cotton socks, is swift to point out that the Sex Pistols are calling the Queen 'a moron'. But hold, mes braves! Not so fast, I implore you. Surely if this were so then the words would be 'Made *her* a moron'! Perhaps the lyrics to 'God Save The Queen' are intended to mean – if they mean anything at all, which is arguable – that a dehumanising society dehumanises everyone, and the words 'the Queen' are used here as they are in numerous stirringly patriotic and utterly wholesome hymns and prayers, anthems and addresses, to symbolise the country as a whole.

That, of course, would make for a much less exciting story (which, I don't doubt, would vex the band as much as the newspapers) and would mean that, rather than assaulting Her Majesty, the Sex Pistols were merely making a comment on over-government, their arguments rather ironically endorsed by the blanket ban on their record.

Their view may not be a popular view (although I challenge you to bring me a citizen who does not object to the intrusion of government into virtually every aspect of our lives), but it is a view that should be allowed to be aired. After all, Enoch Powell, for example, is a man whose views are obnoxious to a great many people, and he is positively encouraged to air them on such programmes as *Any Questions*.

So we are left with only one possible conclusion – that potentially unpopular opinions may be heard only if they are the right opinions being expressed by the right people – and I do mean 'right'. And that is something the formerly Socialist *Sunday Mirror* should really be worrying about. The extent to which people are being misled by the hysterical over-reaction of the paper – and the press generally – to the Sex Pistols is graphically demonstrated in the same issue of the *Sunday Mirror*, where, responding to a *Mirror* photograph of the 'God Save The Queen' T-shirt, an Irene Harris writes, 'I hope whoever had this disgraceful T-shirt printed is punished firmly. They must be Russians.'

See what I mean? See what the Sex Pistols mean? And thanks to Stanley Reynolds of the *Guardian* for being the only newspaper columnist to write rationally about the whole business.

# Half Man Half Biscuit

*Observer*, 26 January 1986

---

'ME MAM DIDN'T KNOW I was in a band until a month ago,' said Nigel Blackwell, rhythm guitarist and singer with Merseyside's Half Man Half Biscuit, teen sensations that are, as we disc-jockeys yell, sweeping the nation.

The band had walked to the pub from Bristol's Tropic Club past a queue which ran down the stairs, out of the door, up the street and around the corner. 'I should be queuing with them,' mused Nigel, faintly embarrassed by this palpable evidence of demi-Biscuitmania.

In the pub, poet turning singer Jeggsy Dodd shuffled paper, selecting a set suitable to the time and the place. His best material includes a piece called 'Wine Bar Man', firmly nailing the brutes with the Dickie Davies haircuts and shirts 'open to the knees', and some musings on Manchester United manager Ron Atkinson.

Half Man Half Biscuit's opening shot was the unequivocal 'It's Fred Titmus', reflections on the feelings engendered by coming face-to-face with celebrities in the supermarket, and a track from the LP *Back in the DHSS*. The eager consumers bellowed the title each time it occurred and some foolhardy spirits at the front slam-danced beneath the perplexed gaze of two bouncers.

'Jesus Christ, come on down!' roared Nigel, introducing '99% of Gargoyles Look Like Bob Todd'. Half Man Half Biscuit's cultural references are spot on, characters drawn from your least favourite television programmes, settings from children's television's *Trumpton* and *Chigley*.

Their lyrics are the first genuinely and consistently funny pop lyrics since Viv Stanshall wrote for the Bonzo Dog Band, and they are set to unexpectedly memorable tunes, the whole rendered in a manner more appropriate to rehearsal than to performance. Somehow this is as it should be.

At the Tropic Club, most of the songs came from the LP or from a session recorded for Radio 1. The latter yielded forth the superb 'Trumpton Riots' and 'All I Want For Christmas Is a Dukla Prague Away Kit'. There were newer songs, too, presumably titles from the impending EP, including an excruciatingly tasteless song about the late Hattie Jacques.

The set ended with the rowdy 'I Hate Nerys Hughes', and the band shuffled awkwardly offstage. A minute later they shuffled awkwardly back again as a section of the crowd chanted, somewhat enigmatically, 'Keith Chegwin, Keith Chegwin,' to perform 'The Trail Of The Lonesome Pine', a title best known in its version by Laurel and Hardy.

It would be difficult to predict the future for Half Man Half Biscuit, for in success could lie the seeds of their own destruction. Perhaps it is enough, as Nigel seemed to think, that they have recorded 'Back In The DHSS'. In Bristol they were a tonic, giving me my best night out in years. You must see them.

# Happy Mondays

*Observer*, 4 December 1988

---

IT IS SOMETIMES PLEASANT, when reading the morning papers, to allow the imagination to skitter forward three or four hundred years and know that by then our football hooligans and drug abusers will, along with plucky little Belgium, have achieved mythic proportion as the only people not prepared to do as they are told by Mrs Thatcher.

From their reviews, Manchester's Happy Mondays seem routinely to be associated with drug abuse, and it is certain that some of them have the bearing of football hooligans. 'You bunch of boring bastards,' someone sang from the stage of London's Dingwalls on Monday night, before going on to accuse us of being Southerners.

Between themselves, however, once onstage, communication seems to cease – to the point at which the songs emerging from the mêlée threaten to collapse in ruins. Yet at the very point of disintegration all is resolved, the numbers thereby taking on a new and different strength. (Jimi Hendrix used to do this. There were times when you swore he would have to stop, admit that he had lost his way in the tumult of his performance and start again. Of course, he never did. Instead, in a trice, he brought order to the chaos – only, as you caught your breath, to sail into another maelstrom.)

Shaun Ryder, one of the great singers-that-can't-really-sing, dislocated eyes filled with scorn, seldom remembers that convention insists that the microphone is the preferred delivery point for the pop vocal. At his side, Mark Day is a man with the rare ability to startle on electric guitar. Against the wall, seraphic, motionless, Paul Davis plays keyboards. All seem unaware of each other's presence, linked only by the homicidal drum and bass of Gary Whelan and Paul Ryder.

About them, wild-eyed men with faces like worn gravel paths dance, mutter, sing and shout abuse. On the floor flailing ragamuffins strive to establish Northern credentials. Twenty-five years ago, pre-poser posers, myself included, pretended to be Scousers. Never in my most fevered imaginings did I think anyone would aspire to being considered Mancunian. Even the Japanese women present – there are always a disproportionate number of Japanese women at the more interesting London gigs – danced, perhaps hoping that they too would be taken for Mancs.

I was too busy enjoying myself to note whether the Happy Mondays were playing selections from the new LP *Bummed*. I expect they were.

# Hello Tailor!

*Punch*, 7 April 1976

---

WITHOUT QUESTION WE accepted that it was neither meet nor seemly that anyone outside the First Eight should wear socks with clocks on. Without a murmur we resisted the urge to wear underclothes that were other than white. Not a voice was raised in protest that we were forbidden to affect skull-caps with tassels and piping in house colours until we had achieved some success in cross-country running. It was my colleague Tony Blackburn who recently urged, on a phone-in programme on BBC Radio London, the re-introduction of capital punishment in schools.

At Shrewsbury we never went quite that far, but I learned painfully that the wages of sin came not singly, but were visited four-fold or six-fold upon the pyjama'd body as you crouched meekly over cupboards containing Cadet Corps uniforms. Retribution came a few minutes after 10 p.m. and the suffering of the beatee was compounded by the knowledge that the whole house lay abed in silence, hoping against hope for a whimper of pain. Not that I ever illegally wore socks with clocks on – I still don't know even what that means – and my underclothing was always whitish. And you couldn't have paid me to wear a skull-cap, with or without tassels and piping in house colours. The sins I had committed were even more obscure than these. Perhaps I had whistled in the passage, forgotten to fill in the changes book, crossed Oliver's Egg. (This latter was a cindered area near the Armoury. I forget exactly which privileged creature was allowed to loiter there, but it wasn't me.)

For the stupid, the unathletic, the unattractive, life was pretty grim. But none of us ever questioned a single one of the demented regulations, nor the system which allowed senior boys rights over their juniors that would have given Simon Legree pause for thought. Yet beneath the layers of indifferently cooked foodstuffs that adorned my jacket, my regulation dark blue jacket, the fires of rebellion were being fanned. We were obliged on Sundays to set aside this dark blue jacket along with the grey trousers and grey socks and wear instead blue socks (clock free) and a dark blue suit. It was at this latter that the revolutionary Peel (then still a Ravenscroft) struck while in the third year of his servitude. During the holidays he visited a gent's outfitters in Liverpool and came away wearing a determined smile and bearing a brown-paper parcel.

On the first Sunday of the new term the contents of this parcel were to be seen across the school site. A person clinging to the rafters of the chapel that morning (and in those days I thought I knew who that might be) would have been able to pick me out from among my six-hundred-plus peers with ease. For the New Suit was light blue, and double-breasted. Not many people, I admit, commented on it, but I knew that I was operating outside the law and the knowledge made me rejoice hugely. Only a tiny mutiny, I concede, but to the sixteen-year-old victim of a hundred small injustices, a turning-point.

# Adrian Henri

*Radio Times*, 20–26 January 2001

---

NOT FOR THE FIRST TIME, my new year's resolution is to make more of a fuss of our friends. Why? Well, within the past fifteen months, Teddy and Peggy, husband and wife, both died and I was too busy on *Home Truths* business to go to either of their funerals, although Sheila did. I have felt twinges of guilt over this ever since, because they were people we loved and Teddy had been responsible for keeping me in work on Radio 1 for years.

Then, just before Christmas, we went to Liverpool. I had been awarded an honorary doctorate by Liverpool University for reasons that were never entirely clear to me. My dad would have been proud of me in a what-has-that-damfool-boy-done-now? sort of way, I thought, and was touched and pleased. William, who had won a degree from the same university through, he maintains, the sweat of his brow, a substance produced in exceedingly limited quantities, affected to be outraged that I could just turn up in Liverpool for a morning and return to East Anglia with a doctorate, but when I checked with him he didn't seem to mind really.

The ceremony took place in the Philharmonic Hall, the very venue in which we had sat to watch William graduate two and a half years ago – with me trying not to cry – and in which, years before, I had introduced a performance by David Bowie to a rapt audience of young persons. Interestingly, the young persons were not there, by and large, to see Bowie. Rather, they were gathered together in the name of Marc Bolan and Tyrannosaurus Rex, shortly to transmute into T. Rex. David was doing his mime act then – with me trying not to laugh – and was bottom of the bill, below an Australian sitar player. I had no idea then, as I have no idea now, of the strength in depth of the sitar-playing tradition in Australia but if I remember correctly, our man made up in sincerity for what he lacked in skill. Next week: the Didgeridoo in Karnataka: a Sideways Glance.

Afterwards we were invited for lunch at the home of the university's vice chancellor and I sat down next to a place-setting in front of which there was no chair. Asking about this, I was told that the poet and painter Adrian Henri – a wheelchair user – was expected. Sheila and I were pleased about this, having spent a fair amount of time with Adrian in the late sixties and early seventies, but we were warned that ill health might keep him from attending either the lunch or the afternoon's ceremony, at which he, too, was due to be honoured with a doctorate.

Our paths had first crossed shortly after I started reading Adrian's poetry, along with that of his fellow Liverpool-based poets Roger McGough and Brian Patten, on the pirate station Radio London. I read it all extremely poorly but they were all too kind to laugh out loud. Firstly I met Andy Roberts, Adrian's friend and accompanist, and eventually the great man himself. Adrian and Andy were in our flat when the Americans first walked on the Moon and they were with us when Sheila and I had our first, brief holiday together in Ireland.

One afternoon, in a place I've never been able to put a name to since but believe to have been a few miles north of Athlone, we walked across meadows filled with wild flowers to picnic in the sunshine beneath the walls of a great castle, sitting on smooth rocks in the shallows at the edge of the lake that surrounded the crumbling fortifications on three sides. On the way across the fields and surrounded by all this faintly excessive loveliness – I originally typed in the words 'the swooning air was heavy with butterflies' but thought better of it – we had stumbled across the corpse of a cow. Adrian had insisted on photographing it from several different angles and in close-up and I was rather appalled. Later I came to understand that that was why he was an artist and poet and I was a DJ.

Over the next few years I did gigs with Adrian and Andy's band, the Liverpool Scene, and am named as the producer of their first LP. The most memorable of these gigs was an anti-nuclear concert in a park in east London. The locals didn't much care for the hippies and protesters in their midst and made their feelings known by pelting the stage with stones, bottles and cans, leaving me, understanding it to be part of my duty as compère, to protect the artistes, hopping about the stage fielding missiles as Adrian and the band soldiered on behind me. They were playing, among other things, their 1969 mini-hit and the dance sensation that, to be honest, wasn't sweeping the nation, 'The Woo Woo'. The East Enders weren't ready to woo woo.

Adrian stayed in Liverpool and we stayed in Stowmarket and we lost touch over the years. But this autumn Sheila and I saw him, briefly, at a birthday party in Heswall. He looked desperately ill, but in that irrational way you do with people you envy for their ability to derive everything there is to be derived from life, we assumed that somehow Adrian would take up his bed and walk and become a force for good all over again. As you know, Adrian didn't revive, he didn't take his place at the vice chancellor's lunch and two days later I heard on the *Today* programme that he had died.

As I said, we should make a fuss of our friends.

# Hippies

*Sounds*, 21 May 1977

---

NOW HERE'S AN INTERESTING one. Cherry Spottiswood of Buxton has written to say, 'Can you tell me, you vision, you Adonis, whatever became of the so-called "Beautiful People" of the late 1960s and early 1970s?'

Yes, Cherry, I believe I can. I was, all joshing aside, always slightly intimidated by the Beautiful People. For a start, I suspected that I was rather too fat to qualify as a Beautiful Person myself, and I always looked like a fairground facsimile of an Eastern potentate when I dressed in velvet and festooned myself with beads.

Spiritually I was right there with them though. I longed to travel to Ranjipore (or wherever it was they all travelled to) and lie for weeks on rich carpets and amid Nepalese cushions get smashed out of my mind. Unfortunately I had to earn a living instead.

I also hoped that when the time came I would have the nerve to call our offspring Canute of Iphgenia or 'Second Thoughts' or Tolkien or Hiawatha. At the same time I had a mental picture of the luckless little brute having to announce himself on his first day at school, and my resolve melted away and he turned out to be called William instead. I think he will be grateful for that.

I had assumed that the Beautiful People had been absorbed

into society, had become accountants and mothers-of-two, provincial art masters and dentists' receptionists. But I discovered only yesterday that they have survived – and as Beautiful People too – in remote rural areas. With a gang of village layabouts and our district minibus I crossed down (or up) to a charming spot called Mettingham Castle, a tumbledown er . . . castle in the badlands of the Norfolk–Suffolk border for the Bungay Jubilee May Horse Fair. And there the buggers were. Hundreds of 'em. A veritable sea of velvet.

Those from the wealthier homes were on horseback, shouting interesting and instructive things between one another for the benefit of passers-by, bumping their meddlesome nags into the grateful peasantry; while their fortunate brethren sold us hideous leather goods, monstrously ugly works-of-art and scruffy dresses.

I had never noticed before how essentially middle class these radiant folk are, either. In the most sublimely cultured tones they explained that, no, there were no toilets, called for their children (these latter mainly blessed with Arthurian names, dressed in gypsy-chic and deserving a sharp boot in the arse) and tried to persuade potential customers of the physical and moral benefits of drinking coffee made from hedge clippings. They sat around fires, alongside beautiful Romany caravans, strummed out-of-tune guitars, and tapped incessantly on 'hand crafted' 'Indian drums', with a sharp disregard for tempo which was doubtless the result of remarkable inner cleanliness.

The Pig, bless her, who, since a whole catalogue of unfavourable experiences with Beautiful Persons around the turn of the decade, has little time for the university-trained peasant surrogate, grew very restless very quickly; an hour later we were back on the road in search of more plebeian amusements.

I hope that answers your questions, Cherry.

# Ipswich

*Disc*, 1970–1

DO YOU EVER, as you drift through life bumping into things, feel that somehow and somewhere a massive switch has been thrown and reality, as we know and have come to love it, has slipped ever so slightly?

Life gives the outward impression that everything is proceeding roughly as planned but you experience, nevertheless, a feeling of unease. A sixth sense tells you that all is not as it seems, that something has gone wrong. It was like that last night in Ipswich.

These days I live no great distance from Ipswich and I wondered why it was that I was instructed to arrive at the Manor House Ballroom as early as 6.30. However, being reluctant to sacrifice my massive fee for being three-and-a-half minutes late, I was there on time.

Those of you who know Ipswich will know also that the city fathers are so keen to keep people out of their city that they have made it impossible to park anywhere within walking distance of the centre of town. However, I found a discreet and hidden double yellow line only a hundred yards from the Ballroom and advanced steadily on the site of the evening's fun and games. The first person I saw told me that I wasn't needed for an hour so I wandered lonely as a cloud – and rather more lovely – through the silent streets.

Having found a rude magazine to read I returned to sit on the step outside and count the customers as they poured in with smiles on their lips and songs in their hearts. At 7.30 I had counted none – if such a thing is possible. A Zen exercise – something like the sound of one hand clapping. Over the next weekend spend an hour or so contemplating the concept of counting to zero.

Clutching a stack of records I strode energetically through the front door and found to my consternation that no one had crept in through the back, nor had anyone been lowered through the ceiling. For forty-five minutes I played a succession of foxtrots and gavottes to the ghosts in the deserted ballroom. It has a strange melancholy – a hint of the Rudolph Frimls – working in an empty hall.

As I stood there shaking my hips to the sounds of the Dead, Alice Cooper, Man and others (who have their debut quadruple LP, recorded live at the Fillmores North, South, East and West, released ever so soon) I forbore to make any announcements. Talking to an empty room was, I considered, rather more eccentric than playing records to it.

Perhaps, thought I, it was all part of an elaborate joke. In a moment, in the twinkling of an eye, merry, laughing people would dash out from the dark corners and recesses in which they were hidden and throng round the table behind which I writhed.

Eventually, of course, one or two people did drift into the ballroom. The rest were, I was assured, in the room upstairs and would be down around nine o'clock. But wasn't nine o'clock, I asked, the time at which I was to be replaced by a band? Er, yes, that's true, someone admitted. Well then, I reasoned, there seemed to be little point in my soldiering on. True, true, everyone agreed.

With a sigh I gathered up my records and went back to Friday (a Land-Rover). Behind me I left a Free LP and a room which seemed to be filling as I left in the way that the garden floods if you remove the bung in the water barrel.

Pig was surprised to see me home at 9.15 – it was the first time I'd ever returned from a gig before dark – and the failure did give me a chance to get tonight's 'Boogie Night' programme together properly.

When I was a kid I always wanted a dog. We lived in the country, about half-way between Chester and Birkenhead, and I was always short of amusing companions. There were the Whittimores about a mile down the road and a little girl in the village who has passed into the family mythology by giving me an adjustable spanner for my third birthday – there is even a photo extant of the ceremony – but I was unpleasant even as a child and kept pretty much to myself. I did want a dog though and for one birthday, just before I slipped into double figures, I had one but my mother wearied of him the following day and gave him to a farmer.

Yesterday I got my first puppy and she is amazing. One of these days she'll be an Old English Sheep-dog but right now she looks like a small and slightly out of control bear. For reasons that are too absurd to explain she has the same name as the cat. They're both called Woggle and at this moment they don't get on very well. Woggle (Dog) often wanders over to Woggle (Cat) to exchange pleasantries – which she does, at the moment, by leaning on you.

The cat isn't totally in sympathy with being leant on and reacts rather negatively by spitting at the puppy, which then wanders away in confusion and registers its bewilderment by first banging her head on some piece of furniture and then performing one of a rather impressive range of evacuations from one or other of the orifices scattered about her body.

More anecdotes and tales of the stars next week.

# Michael Jackson

*Observer*, 17 July 1988

---

THE LAST TIME I saw Michael Jackson at Wembley, he was a diminutive fifth of the Jackson 5, cute and precocious. On Thursday he returned as superhero, larger and apparently stranger than life, with a show I do not expect to see equalled in my lifetime.

The evening paper had a story of spivs selling tickets for the Jackson show at £150 a time. The dispirited scousers I spoke to outside told an altogether different tale. Ruin, I gathered, stared them in the face. Repressing the urge to press coppers into their hands, I continued the yomp in from beyond the extended wheel-clamp tow-away zone, having left the racer in an area which only a year or so ago was probably all sylvan glades and babbling brooks alive with carp.

Once installed in Section 80 with the £5 tour programme and a packet of plain crisps, I settled back in anticipation of a feast of fun. All about me citizens were peering at the empty stage through cardboard opera glasses bearing the 'BAD' logo, whilst others attached 'BAD' balloons to their clothing. Below us the huddled masses cheered each time the Michael Jackson Pepsi advertisement appeared on the screens at the side of the stage.

At six o'clock we cheered as technicians took their places. Five minutes later we enjoyed the first of a few Mexican waves. An hour later Radio 1's Gary Davies appeared to ask whether we were ready to boogie before urging a big Wembley welcome for Kim Wilde.

I felt a bit sorry for Kim. Very much the bread roll with which we toy absent-mindedly while awaiting the meal, she had yet, as the tabloids had emphasised with their usual quiet persistence, to meet Michael Jackson. But there she was, waving a red scarf and bending over a lot so that the cameras could catch the cleavage. 'It's great to be here,' she said. After a song or two a discussion developed in our row about the catering staff, who were dealing out the lager and cold dogs in what seemed to be Motherwell colours. We reached no important conclusions.

In the interval we amused ourselves by leaping up from time to time to gawp at celebrities arriving in the glass-fronted banqueting suite. We liked Frank Bruno best. But suddenly there was thunderous music from the stage, a battery of lights blazed out over the audience and there, scarcely believably, was Michael Jackson.

'How ya doin'?' he asked after a couple of hits. Well, I was as fine as anyone with sore feet standing in a cold, damp football stadium could be – but how was Michael? From close-ups on the twin screens, he did not look too good. The famous remodelled face glowed faintly inhuman beneath a surfeit of rouge – and his performance to date had been curiously uninvolving, despite our overfamiliarity with it from a host of videos.

But Michael Jackson clearly needs a few minutes to get into gear and as the costume changes came and went and the stage and lighting effects grew more audacious, he took control with a performance of matchless virtuosity. Making much of stagecraft learned, surely, from James Brown – especially a device whereby a song apparently finished, with the star seemingly in emotional crisis, frozen save for lips moving as though in prayer, would be reprised – Jackson led his dancers, singers and musicians, all fearsomely well drilled and rakishly handsome, through less a sequence of songs, more a series of scenes, the whole resembling some futuristic, technological pantomime, with Michael Jackson himself a distillation of all principal boys, singing some of the world's best known songs and dancing with such authority, timing and energy that the odd action replay would not have come amiss.

My only wish is that my children could have been there to see this stupendous performance. It is something they would never have forgotten.

# Billy Joel

*Observer*, 12 July 1987

THE LEADER OF the drove of squat Americans outside Wembley could not understand why possession of a particular credit card did not entitle her to go to the head of the queue for tickets. Her party had flown over just for this series of concerts and she expected better treatment than this. 'England,' she explained to her henchwomen, and they chortled conspiratorially. I was never so proud of being English.

Inside the Arena, £10 bought a Billy Joel T-shirt or two-and-a-half programmes. The programme was a special tour edition of what I take to be the regular Billy Joel fan magazine. It is called *Root Beer Rag*. It would be. It just would be. The editor of *Root Beer Rag*, writing of the latest Joel LP, gushed that *The Bridge* 'resonates more than ever with diversity and expansiveness'. Does that mean anything? Answers on a postcard, please.

The concert started, after a twenty-minute delay during which the audience applauded both the playing of 'A Hard Day's Night' and a puff of smoke from the dry-ice machine, with a tape of one of those mock-classical pieces that once illumined *Family Favourites*. It was far too loud and the audience clapped along. When the lights went on, we saw a bare stage with a range of keyboards scattered over it and heard a well-drilled and severely routine rock band, its members dressed like hip missionaries.

The only splash of colour came in the yellow shirt of a synthesizer operative whose melancholy task it was to play the role of crazy young guy who was so excited, that, you know, wow, he just could not contain himself. I disliked him intensely.

Mind you, I liked the saxophonist even less. His job was to cheerlead and, gosh, sometimes play percussion and yet another keyboard *at the same time*. He was given to pointing at imaginary acquaintances in the front rows and shouting things that were plainly inaudible. He was a regular guy.

But even this horror paled into nothingness when compared with the chap sitting next to me. As Billy Joel moved from ponderous song to ponderous song, each top heavy with florid pianistics and mock heroic lyrics, my neighbour clapped along, driven by some lunatic inner rhythm.

Below me the faces of the American trippers shone with religious ecstasy as Billy ground on. The band swapped instruments, clowned, were regular guys, laid down some baaad guitar and clapped their hands over their heads. 'Good evening, London,' joked Billy. Everyone cheered. He played 'Rule Britannia'. Everyone sang along, then cheered. He mentioned his last tour here, three years ago. Everyone cheered '1984'.

He sang that horrid song about the piano man. You know, the one with the tonic and gin line that sets your teeth on edge. When Billy started pretending that the low key but obviously necessary security was some sort of proof of British indiscipline, I could take it no longer.

'Try to imagine,' I said to the people in the Tonibell van outside, 'Elton John without the costume, the sense of the preposterous or the tunes.' 'That bad?' one of them asked. It certainly was.

# Kerguelen

*Radio Times*, 16–22 December 2000

---

I THINK I HAVE mentioned before that *Scorn*, described as 'a bucketful of disparagement, invective etc.', all chosen by Matthew Parris, never leaves our bedside. It is a compendium of the sort of things you wish you had both the courage and wit to say to or about people. On the back cover there is a portrait of the compiler, smiling. The smile looks a little unnatural, even slightly sinister. It is the smile of a man who would like you to believe that he is more confident than he actually is. It reminds me of early Radio 1 publicity shots of Tony Blackburn.

Now Parris, former MP, sketchwriter, broadcaster, compiler, critic and more, has had enough, in his words, of 'sneering for a living', and is to undertake a journey so filled with potential hazard, so certain to be seriously uncomfortable, that most of us would throw a log on the fire, snuggle into the corner of the sofa and shiver at the very thought of it. Matthew – 'How I loathe the sight of myself trying to be clever on TV' – is to go to Kerguelen for *To the Ends of the Earth: Dreaming on Desolation Island*. Kerguelen, I hear you cry. Where on earth is that? Well, it is 2,000 miles south of Réunion, and a sea journey there – the only journey possible, as the weather is too extreme for flight – takes you through the winds of the 'Roaring Forties'. You'll remember from school that the 'Roaring Forties' are best avoided. In our

*Times Atlas of the World*, Kerguelen is on the very last page, page 179, before you get to the index. Even then it is a mere inset on a page devoted to Antarctica. That's how remote Kerguelen is. About the only thing dull or predictable about Kerguelen is that it is the length of Wales, that universal measure of area and distance for places that are not the size of Belgium. (Channel 4 publicists claim Kerguelen is the size of Cyprus, but I think they're just trying to be different.) Foul weather, snow, hail, sleet and howling winds are a constant on the island, as are penguins and French scientists. Kerguelen is French, but no one cares. As there isn't a soul on the island apart from the scientists, there is no liberation movement.

Matthew Parris has been fascinated by Kerguelen since he was a boy and has now been there – without the reassurance of a film crew shivering out of shot behind the camera – and tells us what he found. He seems, understandably, rather chastened by his experiences, the early sneering – 'Penguins are very New Labour . . . angry . . . suspicious eyes' – replaced by humility and something akin to awe.

Kerguelen is so called because it was discovered in 1772 by the French navigator Yves de Kerguélen-Trémarec. Cap'n Cook happened upon it four years later: he preferred Desolation Island, but this slightly B-picture alternative never really caught on.

Matthew stays with the scientific community of fifty-six Frenchmen and two Frenchwomen – he has little choice to be honest – despite having little French himself and worrying about their reaction, should they realise that he is gay. He spends four months with them, during which he films himself going on a hunting trip into the gale-blasted interior – he is saddened, as I would be, when the hunters kill a reindeer – circumnavigating the 6,000ft Mount Ross and trekking to the abandoned whaling station at Port Jeanne d'Arc. This whaling station was abandoned by its Norwegian occupants in 1929 and, as you know, it takes a lot to make your Norwegian abandon anything. The reindeer were introduced to the island by humans, as were the feral cats that live in rabbit warrens, although the wingless butterflies arrived there as the consequence of one of nature's little jokes and I imagine have been livid about it ever since. They are wingless because there is no point, in that climate, in even considering coquettish fluttering. Parris also leads us to the remnants of the pitiful settlement that marks a French government attempt to introduce sheep-farming to the benighted island. There was a time, apparently, when dinosaurs roamed a richly forested Kerguelen, but those days have gone, possibly for ever.

Parris is concerned, on these trips away from the scientific/military base, that his companions are in their twenties, whereas he is fifty and his toe hurts. But he keeps going and we should be grateful that he does, because he gives us the only insights we are ever likely to get into this monstrous place. Kerguelen is, of course, beautiful, albeit in a terrible way – there are loads of birds, including enormous skuas, and there are sea lions, or what I think of as sea lions. They have, as is traditional in such places, no fear of man. As you look at these creatures carrying on with their business as blizzards rage about them, you wonder what they'd make of, say, East Anglia. It'd be like St Lucia or Bermuda to them.

Before Parris leaves the island, tragedy visits the scientific community. A man is accidentally shot dead. This incident obviously affects everyone, including the former MP. He is compelled to suppress, without much regret, I suspect, his journalistic instincts. He also concludes that, contrary to his former leader's celebrated dictum, there is such a thing as society.

Sometimes, if you'll forgive the cliché, we have to live through hell to learn the simplest of truths.

# King Arthur

*Radio Times*, 18–24 June 1994

---

WAS IT LAST WEEK or the week before that top scientists were elbowing each other out of the way to appear on our radios and televisions to warn us, in voices throbbing with emotion, that global warming meant that we were going to have to adjust pretty smartish to living in a Mediterranean climate? Olive groves, pedalos, armed policemen and corrupt politicians lurked just around the corner, they suggested. Whatever happened to all of that?

This week's fearless analysis of your fave programmes comes to you from a rain-drenched Isle of Man, where seven or eight of us have gathered to eat and drink too much and watch wiry men – and a couple of wiry women – fling themselves about on motorbikes. So far, we have seen precious little racing but have experienced enough precipitation to persuade us that the Manx language probably boasts thirty-two different words for rain.

But, as you know, the words Positive Thinking are tattooed on my knuckles and, despite the fact that I have a heavy cold and have just banged my head, I have turned the wretched weather to my advantage and watched a lot of television. Sadly, reception here in Kirk Michael is not too good so I have been forced to watch Ceefax most of the time, especially the Foreign Office guides, country by country, to travel overseas. If you ever feel that life isn't so bad after all and that the world is by and large, as I seem to remember Blue Mink or someone equally frightful suggesting, just a great big onion, try Ceefax as an antidote. Banditry, mugging, thuggery and riot seem likely to engulf you wherever you go and whenever I am sent to the village for yet more red wine, I go prepared for mayhem. So far all I have encountered has been the odd enormous German with a daft moustache who blocks the pavement and forces you to step into the road to get past. Innocent enough fun, I suppose.

One of the many programmes I have not been able to enjoy here has been *Spitting Image*, but I am told that there is now a puppet of Peter Beardsley. I am a great admirer of Beardsley's skills and the closest I have ever come to football violence was when he played at Carrow Road, Norwich, a few seasons ago. Beardsley played at the time for Liverpool, before the club entered what I like to think of as their conceptual period and replaced most of their good players with items of furniture apparently selected at random, and whenever he addressed the ball an oaf behind us sang, 'Ugly man, ugly man, ugly man'. Needless to say, the singer himself looked as though a playful providence had modelled him on the Pompidou Centre, with all his facial plumbing on the outside, but he appeared oblivious to the irony in his weedy chanting.

You will have read elsewhere of the man, featured in ITV's *3-D*, who believes himself to be the reincarnation of King Arthur. Back in the hippy days of the late 1960s you could ride a mettlesome nag full seven leagues before you encountered a man who didn't believe himself to be the reincarnation of King Arthur and if you went, say, to Stonehenge or Glastonbury you found your way blocked by hordes of gibbering former Kings Arthur. I kept the jawbone of an ass in my briefcase specifically for the purpose of smiting them hip and thigh on such occasions. Actually, why is it that anyone who claims a previous incarnation has been someone really interesting? We can't all have been Kings Arthur or Henry VIIIs or Queens Guinevere or Cleopatra. I myself was previously a six-mile section of the East Lancs Road and before that, Jim the cabin boy on a tea clipper plying her trade out of Plymouth. I died in a fall from the very tip-top of one of those mast things as we rounded the Horn in a nor'-nor'-easter.

Outside, the rain continues to fall. My only consolation is that it will almost certainly fall on Wimbledon as well. There is something about Wimbledon that brings out the worst in me. Not sure what it is really but the blanket coverage, the fawning commentaries and the behaviour of the stars, behaviour that would have got them boiled in oil in a decently ordered society, play their part. The only Wimbledon I enjoyed was the one in which torrential rain resulted in flash flooding and chaos. I hardly missed a moment, a spiteful smirk on my spiteful little face. Fingers crossed, eh?

# Knebworth

*Sounds*, July 1975

---

I WAS TWO HOURS late for my meeting with John Peel. I had made arrangements with Viper, his Polynesian secretary, to meet with John in his elegant Pimlico pied-à-terre, but I had been caught in a considerable 'confiture-de-traffic' and arrived on his spotless white doorstep, breathless and without my Hokkaido-San TJ 45/2a portable recorder, several minutes after noon.

I experienced a measure of difficulty before I was actually allowed onto the doorstep, the officer in charge of the squad holding the crowds of young girls at bay refused to believe that I was there on business, and it was only when Viper phoned down from the penthouse flat to confirm that I had an appointment with John that I was allowed in.

After the dust and heat of the street the front-hall was a haven of peace and tranquillity. As my eyes became accustomed to the gloom I saw some of the trappings with which a fantastically successful DJ surrounds himself. In a corner of the vast hall was parked a 1925 Hispano-Suiza, on which generations of fans had scrawled their protestations of unfailing devotion in a bewildering variety of cheap lipsticks.

On the wall was a range of fine paintings. I recognised a Fibolini, two de Rundfunks and at least five Scherpelmeters. There was also a full-sized map of the Paris Metro system and several posters advertising John's sensational gigs in places as far apart as Birmingham and Coventry.

I was allowed to linger in this attractive environment for two and a half hours – unfortunately there were no chairs in sight – before Viper, a stunningly lovely creature with a *South Pacific*-styled coiffure and a tartan sari of fetching simplicity, came down to tell me that I could now go down into the kitchen and fix the refrigerator.

I explained that I had been sent by *Sounds* to interview John and to discover his feelings on the Knebworth concert. She laughed and disappeared upstairs again.

Forty minutes later she reappeared, as lovely as ever, and told me that John would see me. The lift didn't seem to be working, so Viper and I walked the eight floors up to John's private suite.

In between gasps I asked her what it was like to work for such a fabulous personality. Was it true, I wanted to know, that he was making a film based on the life of Gibbon, the personable author of the definitive work on the rise and fall of the Roman Empire? Had he, I wondered, really severed his relationship with glamorous starlet Tricity Vendome? Viper passed me a grapefruit from a bowl on the staircase but made no answer.

From time to time as we climbed I saw the shadowy and scarcely formed figures of graceful young girls of all hues and nationalities flitting in and, indeed, out of such rooms as the library, with its unique collection of rare illuminated manuscripts, the billiard room, with its teak-panelled sauna and massage chamber for the exhausted billiardier, the jester's gallery, with an extensive Hornby-Dublo layout, and the private zoo, in which John keeps the three white panthers he takes with him when he tours the clubs and discos of the West End.

Eventually Viper and I reached the great man's suite and were ushered in by the two dusky dwarves, who come, I believe, from Dahomey.

John was already up when I walked into the room, and one of

the girls in the room offered me a Tequila Sunrise. I accepted gratefully and handed her the nutritious citrus fruit Viper had given me five minutes earlier.

John beckoned me over to the brushed-steel and glass bar beside the bed, motioned me on to a Victorian rocking-horse adjacent to it, and clambered back into bed. A furtive giggle came from what I had hitherto supposed to be a heap of clothing on the pillows. Having made my apologies for my tardiness, I began the interview.

*Sounds*. Er, John, do you feel that, in the light of present day developments, and with regard to the generally deteriorating economic situation, it is important – or indeed relevant – that many thousands of bewildered and disorientated young people should gather in a field in Hertfordshire for an event which, admirable though it may be, can only serve to foster the impression that the rock and roll industry – if I may call it that – with its international ramifications and connections, and with the rumours of large-scale involvement by organised crime syndicates, is embarking on what we can only call – as Van Kuyper did in his essential treatise 'Der Volken von der Vareweg Trie' – the 'blood-rush of ultimate decay'?

And should such established and, dare I say it, powerful figures as yourself allow impoverished young people of the proletariat to devour their substance in riotous living, as it were, by paying hard-earned money to see you making announcements to the effect that would Mary and Denise meet Rob and Gunk beneath the Coca-Cola sign, please, across several hundred yards of greensward that could more effectively be used to grow such high-yield crops as celery and Etruscan corn?

My only answer was a snore.

# Kosmische Musik

*The Listener*, 12 April 1973

'MUSIC IS TO GO through acoustic irritation. To make music is the reflection of acoustic environments and their change into special subjective sound patterns which originate from the confrontation between the individium and society. Music is everything which produces sound waves and consequently let swing the air in certain interval (frequent). Frequency is the primitive foundation of music. A particular tone. Several frequents which sound at the same time make a swelling. So, swelling is the totality from every musician possibility, the total of every sound.'

So says Klaus Schulze and he should know, for Klaus is the grandfather, if a man born in 1947 can be grandfather of anything, of Kosmische Musik. His statement, which probably sounded better in German (I can't believe he really intended to say 'swelling'), makes up part of a barrage of documents in fractured English which originate from Berlin, the home of Kosmische Musik.

There's a record devoted to Kosmische Musik on the Ohr label (available spasmodically from fashionable import shops) and Klaus is featured on that record. Also featured are Popul Vuh, Ash Ra Tempel and Tangerine Dream. For anyone claiming an interest in the development of contemporary rock music this double album is essential. Actually, to call the music 'rock' is inaccurate. Rock music really means music based on an Anglo-American tradition – rhythm 'n' blues mainly – and Kosmische Musik owes no debt to any part of the established rock order. The only British band to have clearly influenced the Germans are the Pink Floyd, and Tangerine Dream and the others have extended the spacey music the Floyd pioneered into something quite unique. Klaus Schulze is extensively involved with quadrophonics. He has recorded a semi-formal piece of his own invention with a full symphony orchestra. He has managed to tame the ubiquitous synthesizer and has contrived to make real music with it rather than using it to produce mildly diverting ornamentation for music based on the traditional guitar, keyboards, bass and drums quartet. He was with Tangerine Dream when they recorded their first LP in 1970.

For my money, Tangerine Dream are the best of the Kosmische Musik bands. Whenever any of their extended works are played on the radio there is a heavy mail from listeners. Most of the letter-writers are for it, those that are against it are very against it indeed. A Tangerine Dream track, heard superficially, is little more than a repetitive drone. Closer listening reveals a constantly shifting and evolving pattern – something like Terry Riley's *In* C. It is relevant that these Kosmische groups seem to feel some affinity, judging from LP sleeves, with the work of M. C. Escher. His drawings, in which various shapes are gradually and imperceptibly transmuted into their antitheses, have a lot in common with Kosmische Musik.

On the Tangerine Dream double LP *Zeit*, the instrumentation includes three synthesizers, four cellos, 'generator', steel guitar, vibes, organ and cymbal. An excellent duo, Cluster, who are not officially a Kosmische Musik group but certainly sound like one, use two synthesizers and other instruments only incidentally. Ash Ra Tempel, who have recorded recently with Timothy Leary, also had Klaus Schulze as a member when they first recorded. Their manifesto says: 'We want produce living electronic and use a pure mechanical tone reproduction from a mathematical amorphous apparatus which forbids the personal and impulsive identification from the author with his product.' Despite this disclaimer, Kosmische Musik is very introverted and personal. As Florian Friche of Popul Vuh said (approximately), 'Let's make a sound to lead us from the outside to the inside.' Incidentally, the near-English of the press hand-outs reached a glorious peak when it was stated that in the months prior to their first LP release Ash Ra Tempel played to 'nearly nobody'. That Zen-like concept is almost as intriguing as the music itself. With these Kosmische Musik performers, their spin-offs and the other important bands such as Amon Duul II, Can, the mysterious Faust and Neu, it would not be stretching the truth to say that the most interesting and genuinely progressive music anywhere in the world is coming from Germany.

# Frankie Laine

*Observer*, 1 May 1988

---

UNLESS YOU COUNT a performance by the Obernkirchen Children's Choir, the first concert I ever witnessed starred Frankie Laine, The Voice of your Choice. This was in 1952 at the Liverpool Empire and I went with my mother. Earlier that year I had bought my first Laine record, 'High Noon', and established a pattern that has lasted to the present day by preferring the 'B' side, 'Rock Of Gibraltar'. Much of my pocket money over the next three years, until Presley came along and changed life not so much overnight as over *Family Favourites*, went into accumulating the complete works of Frankie Laine.

This, perforce, involved moving back in time as well as forward, as far back as 1947 and the Italian-American's first hit, 'That's My Desire'. En route were uncovered such gems as 'Chow Willy' – 'Johnny, will you buy me a weddin' gown?' – which Laine sang with Jo Stafford, and 'Girl In The Wood', which rested in my collection alongside the chart biggies 'Answer Me' (banned for its references to the 'Lawd'), 'Hey Joe' – 'Where d'ya get that jolly dolly?' – and the matchless 'Kid's Last Fight' – 'Said the Tiger, scowlin', "Doncha know ya through?"' – from 1954.

Backed at the Empire by Vic Lewis and His Orchestra, Frankie sang well up to expectations but was a disappointment to gaze upon. I knew from a popular publicity shot that, overlit in the spirit of the age, made my man look like an over-ripe frog, that Laine was, even in 1952, no longer in the first raptures of youth, but I had hoped for someone leaner and meaner. I still argue that the light blue, shot silk suit was ill-chosen.

Experience has taught that it is commonly a mistake to attend upon former heroes and heroines as they reprise their hits. Either they elide the songs into a medley and sing them too fast and too apologetically, as Diana Ross did, or run through them in a perfunctory manner with a nervous pick-up band, as Chuck Berry does.

The only exception to the above I have seen has been the celestial Roy Orbison who, two years ago in Ipswich, opened with a mildly untidy 'Only The Lonely', then, despite musicians who looked like the prime exhibits in some nightmarish National Museum of Haircuts and had opened their account with eight minutes of what they plainly – and mistakenly – perceived as evocative space noises, sang with such fervour his catalogue of losers' songs that by the time he reached 'Running Scared', my casual evening wear was awash with tears.

For this reason I am wary of belly-flopping into nostalgia by going to see Frankie Laine, now seventy-five, on his present tour. He is in Birmingham tonight but on the 28th of the month will play the Liverpool Empire again and the temptation to be there is great. I may even take my mother.

# Liverpool

*Disc*, 12 May 1973

GOOD MORROW, fellahin, your lovely Uncle John isn't feeling that neat today. In fact, I seem to have some sort of irregular temperature but will soldier on regardless, displaying the sort of courage that enabled Simon d'Escargot-Peyle to be the first man to set foot on Saxon soil when the Conqueror landed.

The Escargot-Peyles (Peyle became Peel and we dropped the Escargot – and the shell broke right across) were granted the rights to much of what is now known colloquially as Middlesex, but, it's alright, I don't intend to claim it back just yet.

I planned, as I always do, to bring you lurid tales of the multisexual world of rock-a-boogie, but every event of the past week, the breakfasts with David Cassidy and the night at the pictures with Carly Simon included, pales into the merest insignificance when compared with the affairs of last Saturday – the latter having passed already into family legend under the title of 'The Affairs Of Last Saturday'.

When the film is made Steve McQueen will be playing the part of J. Peel, my brothers will be played by the Temptations and my grandmother by Marlon Brando.

On Friday night the icy east winds blew ragged newspapers and soiled paper-cups through the dim streets of Hull. (This is my attempt at fine descriptive writing so that I can apply for a job with one or other of the Sunday colour supplements and get discounts on all that wonderful collapsible furniture – I thought you should know.)

Upstairs in a dimly lit room (that's two 'dims' within hailing distance of one another, but they'll never notice) stood a handsome young man, his firm features thrown into relief by whatever it is that throws one's features into relief, every fibre of his being the cynosure of every female eye in the room.

(I'm sorry about all these brackets, but I cannot keep secrets from you – I don't know what 'cynosure' means, but it's the sort of word you need to use to write for those damned colour supplements. Another useful one is 'cymotrichous' which means wavy-haired. Don't forget it because I shall be asking questions next week.) Now, where was I?

With his strong yet elegant hands he placed record after record on the twin turntables to the sighs of the assembled nymphettes.

When I had finished delighting the throng at the disco I turned Friday's nose towards Bradford and a night's sleep.

In the morning, the Pig and I, together with Phil and Janie (or, as their friends have nicknamed them, Janie and Phil), set sail for Liverpool in a brisk nor'-nor'-wester. In a tavern, we met up with a parcel of rascally fellows which included a certain Ray Kane and the excellent writer/singer Jimmy Stevens, whose *Don't Freak Me Out* LP on Atlantic you have almost certainly ignored – to your cost.

And why had we all come together in that place and why were we later standing with in excess of 50,000 other citizens and why was the foolish Peel crying?

Because Liverpool won the League – just as I told you they would at the beginning of the season.

# Local Radio

*Radio Times*, 22–28 October 1994

---

THIS WEEK I had to drive from East Anglia to Glasgow and decided to use my time profitably by listening to pop on FM as I went. The only condition I imposed on myself was that I would change stations as soon as I heard that Cyndi Lauper record. As I pulled out of our tree-lined driveway (just kidding), I tuned to SGR FM. SGR stands for, I think, Simply Great Radio and their Mr Morning Man in Ipswich was Dave Hoffman. Dave is strictly old school, with that Gary Davies love hiccup in his voice. There are, I was to find, so many DJs still doing this that music-loving space critters landing in Britain would assume Gary was some sort of god to the rest of us. Mind you, they would also assume, from radio advertising, that the population was in the grip of a frenzy of carpet buying and window replacement.

Having left SGR when Cyndi sang, I conjured up KL FM, the New Sound in West Norfolk. Someone-in-the-morning was telling us about Lonnie in New Orleans who had castrated himself with a chainsaw. This is what the kids want, I thought, cranking up the volume, but a succession of bland records with keyboard intros, hissing syndrums and greetings-card lyrics soon drove me away.

Q103 FM has – and I quote – the better music mix for Cambridgeshire. Quite how this is measured, I am not sure, but virtually every station heard during my run claimed to be playing either better, or the best, music. Cyndi Lauper once again drove me into someone else's arms.

Grantham is apparently all of a dither over the arrival of the new range of diesel-powered Fords, or so Lincs FM suggested. Why is radio advertising so pedestrian and, when it attempts comedy, so unfunny? By now it was mid-morning and time for a lively mix of music 'n' chat. The chat on Lincs FM was of poetry (it was National Poetry Day) and of a Mr Pippin who complained about 'lady drivers'. When I switched to BBC Radio Nottingham I got a mystery voice competition and the Carpenters. I turned off and listened instead to the surface noise road that scientists have thoughtfully provided south of the Trent.

Our space critters would assume, that, judging by the records playing during the day, either life had come to an end in 1980 or the ultimate human dream was to get it together with the only one worth thinking of, with a view to making it through the night.

Viking FM told us, several times, of the betrothal of Jackie and Alan, both of Goole, before warning of their intention to make the morning air hideous with the sound of Curiosity Killed the Cat. Birthday Files, From This Day in History and an A-to-Z of Hits were on offer from Humberside and Gateshead and I learned that the Best Music from West Yorkshire is an old record by the Eurythmics. Signal in Cheshire is the 'first, the best' – but at what they did not say.

By now I was growing irritable and not even Andy Peebles on BBC Radio Lancashire with consumer advice could cheer me. After resting, I turned north on the M61 and the situation improved somewhat. JFM in Manchester played me some modern rhythm 'n' blues, including a song about the big-legged woman seemingly central to the cultural life of Black America in the thirties, forties and fifties, but blotted their copybook with talk of real music, a concept as meaningless as best music. Rock

FM was not, disappointingly, a lively mix of guitars and Satanism, but I stayed with them until the Lakes, where the scenery was so beautiful that the radio went off.

When it went back on again, Radio Borders told me where in Galashiels to find vegetarian haggis and West Sound and Jenny-At-Drivetime played me three good (if elderly) records in a row.

If you are one of those who has deserted Radio 1 in favour of trite radio with play-safe music policies, well, there's plenty to please you out there. Otherwise, drive from, say, Preston to Glasgow a lot.

# Lovelace

## At the 'Talk of the Town'

*Disc*, 1970–1

---

IN THE SUDDEN hush the floor rises slowly and on it are all these tall women in high, wide, blonde wigs. Below these, and above what I can only describe as ensembles of sequinned jock-straps and bras blurred behind swirls of chiffon, are grins. The male dancers in their sharp, violet suits grin too. The male singer, in a similar suit, does it light opera style and grins even as he sings.

Nudging one another in anticipation the audience fill their glasses from the jugs of ale or the bottles of Mateus Rose on the tables beside them and lean back, well-fed and contented, to watch the show.

John Walters, his Helen and the Pig and I are in a party at 'The Talk of the Town' to see Lovelace Watkins because Lovelace Watkins is worth seeing for the way he mesmerises the audience with his laughter, sentiment and old-time show-biz panache.

'Dance with me on a wonderful evening like this,' someone sings and more women appear – this time in shocking ginger wigs. The chiffon swirls again and they have vast Victorian bustles behind them. White gloves reach up their arms almost to the elbow, still they grin coquettishly. 'Bernard Delfont presents . . .' and there are even more of what would surely be described as 'long-legged lovelies'. This time they have high, high headdresses bursting into a flurry of simulated feathers. I'm glad they don't use real ones any more.

'C'est Magnifique,' they sing and, in a way, it is. The Pig points out that she'd not be able to stand the pace for five minutes and they've already been going for fifteen – and they're still smiling. It's orange wigs and boaters for 'My Cherie Amour' and then there's a long-legged lovely on the bed – the bed? I don't even know where it came from but it's suddenly there.

Two bowler-hatted gents caress the bed-borne lady antiseptically while four others spin the entire group round at a dizzying speed. On the rotating bed she sings and smiles while the gents nuzzle closer. It can't have been easy.

Suddenly it flashes to me that all the women look, in their white wigs, like my Aunt Joan did when I was a kid and I was always terrified of her – I still am to tell you the truth. I'd be very frightened of the 'Talk of the Town's' dancing ladies.

We'd eaten too much really – the food was fine and the waiter was actively friendly. I think the suit I bought for the occasion did that, because usually waiters are less than amiable with me. The vegetarian angle foxed him briefly but he came up with a plate full of nice things like asparagus and avocado.

An oriental routine opened inexplicably with the theme music from *Shaft* and re-aligned itself rather more logically with 'Night of Oriental Splendour'. The couple behind us, who'd introduced themselves earlier with 'We're from Cleveland, Ohio, and we sure like the way you talk,' were clapping their hands gleefully by now and the dancers, still smiling, were wearing costumes that would not have been amiss at the '24-Hour Technicolor Dream' in 1967.

The dancers wear pagoda type headdresses and, stone me, if six more of them don't come from the ceiling in a mock Taj Mahal of the type you see outside the better Indian restaurants in Leeds or Manchester. The swirl, energy and numerical strength of the cast cannot but impress you. I thought back often to the pantomimes of my youth at the Empire or Royal Court theatres in Liverpool and if Widow Twanky had come on to lead us through 'Fuzzy Wuzzy was a bear' then, what the hell, I'd have been knocked out. The scene shifts slightly from the Far East to the Middle East and we have 'Baubles, Bangles And Beads' and, I think, 'Love Is A Many-Splendoured Thing'. 'Splendour' is a key word here.

Suddenly the visions of the mysterious East vanish and we're on the Left Bank. Slinky women shift about in red PVC raincoats – the lads are still in violet suits although I suspect they're different violet suits. 'Diamonds Are A Girl's Best Friend', 'Diamonds Are Forever' and then it's all over with 'Ladies and Gentlemen, Les Girls' and they sink back into the floor from which they sprang. Ye gods, they do work hard and deserve the applause they get from the blue-rinses and their old men clustered round the stage.

There's a pause for our adrenalin to sort itself out and then 'Ladies and Gentlemen, the Talk of the Town is proud . . .' and Lovelace is there and straight into 'Once In A Lifetime'. He's looking pretty good in a light grey suit with wide lapels. It's difficult to say why he goes down so well – perhaps it's because he really looks as though he was being discovered all over again every night; perhaps it's because he's every bored suburban housewife's fantasy male who's going to jump in through the kitchen window and give her everything the old man tired of giving her years ago. We're a pretty cool audience and he's having to work even harder than usual but he tells us what an incredible audience we are anyway.

'Welcome, it's my pleasure, not too fast, not too fast,' and that high laugh. How can you resist the man? It would be churlish to try. 'Everybody happy? I'll change that,' and again the laugh. 'Come on, everybody, snap your fingers, everybody who likes sex snap your fingers.' 'Ooooooh,' go the blue-rinses and look in mock-shock at the others on their table.

Lovelace gives 'Fire And Rain' a treatment that would astonish you – he even makes it sexy. Then it's 'Every time I sing this song strange things happen – devastating – provocative – demanding.' It's 'Hey, Jude'!

'You'd better get out of here while you still have the chance' and the tables chortle and look eager. On the chorus he holds the microphone for members of the audience who sing out hopelessly but joyfully out of tune with a courage I could never muster. A middle-aged gentleman is hauled up on stage and struts around with the mike as though to the manner born. Lovelace goes and sits in his seat and everyone squeals with delight. The middle-aged gent starts a speech about this man being the 'greatest entertainer in the world' and Lovelace is up on stage again leading the applause and reaching for the microphone before the speech becomes an embarrassment.

He knows exactly what to do with the audience now, plays on them like you might play on a piano – and they love him for it. Here's a tribute to Judy Garland, 'misunderstood, broken-hearted – she can't sing it for you tonight but, Judy, I'll sing it for you', and it's 'Somewhere Over The Rainbow'. The once-cool audience is following him wherever he goes now and he switches the mood from the hopelessly maudlin to the fun-riot with a floppy hat and a Gilbert O'Sullivan song, 'My Friend, Gilbert O'Sullivan'. The lights at the back of the stage go up and the whole orchestra is wearing floppy hats too – more squeals of delight.

As the Pig says: 'He'll do it!' off comes the tie – 'You Make Me So Very Happy'. 'Spinning Wheel', off comes the coat and he ripples his muscles – he's got a lot of muscles and no fat. They love it. 'You've heard me – but I'm going on' and again that laugh.

So it goes on, with the audience completely sold and Lovelace to all appearances having a ball – he probably is too. He mops his brow with a napkin and returns it to the woman on the front table who giggles and puts it into her bag. He sings 'Down On The Corner' – 'Come on, let's have some happy music,' and

'You've Got A Friend' and hitch-hikes his way around the stage.

He sings his encore without the microphone, face grotesqued with the effort of it and with the sweat streaming from his face. It doesn't matter that he doesn't have the best voice in the world, it wouldn't matter much if he didn't sing at all.

He carries the whole thing with every trick in the show-biz bag, like a medicine-show barker, and it's an incredible thing to watch. The folks from Cleveland, Ohio, tell us as they leave that they've seen Sinatra, Sammy Davis, Tony Bennett – the lot – but that old Lovelace here makes them look sick. He does at that.

# Madonna

## How Madonna Dealt with a Hostile Press

*Observer*, 9 March 1986

---

'THERE'S THE CHAP who was knocked down by Madonna's car at Heathrow,' someone said. 'He got up then lay down again for the photographs,' another voice alleged. There has been much publicity, most of it strident and hostile, surrounding the visit to our lovely country of Madonna and her actor husband, Sean Penn, here to complete work on the film *Shanghai Surprise*. It was ostensibly to discuss this film that we were gathered together in a cocktail bar off Kensington High Street where a Shanghai Surprise could well be on the menu – 'a creamy mix of light and dark rums, with a heady dash of bitters and a breath of blue Curaçao. £6'.

The film has been made by Hand Made Films, Prop. Geo. Harrison, and a representative said that they were disturbed by the publicity their stars had received. This may well sound like a goalkeeper complaining that his life would be idyllic if only sweaty men with foolish haircuts would stop kicking footballs at him, but the popular prints have been particularly harsh on Madonna and Sean Penn, principally, one suspects, because they are young, attractive, wealthy, probably having a better time than most of us, and reluctant to involve themselves in the imbecilities expected of celebrities visiting Britain.

This was my second Press conference. The first was in Minneapolis in 1965 when the Beatles fielded a series of questions of great fatuity, including one from me that will forever remain locked in my memory, unshared and unloved. My question then – and I felt myself die a little as I asked it – was directed at George Harrison and it was this same Harrison who controlled the Madonna Press conference.

The two notables sat side by side at a small table. On a yet smaller table in front of them cassette recorders had been laid – no broadcast-quality recording was permitted – in a strange little ceremony. George Harrison, looking fit and relaxed, chewed gum incessantly, very much the international businessman. The Liverpool accent has softened, his voice taking on a slightly American quality.

Madonna was simply, even austerely dressed, pale, beautifully made-up but with her hair sufficiently in disarray to reassure those of us who were mightily impressed with her performance in *Desperately Seeking Susan*. When she smiled she looked vulnerable but defiant.

After George had welcomed us and told us that he wanted a White House-style Press conference, Madonna was asked whether she had ever been a Beatlemaniac, the first of a wide range of daft queries. She fielded this, as she fielded the others, with some charm, answering with apparent seriousness all but the plainly unanswerable. She was, she pointed out, a trifle young to have been much of a Beatlemaniac.

There was, despite this gentle beginning, a palpable hostility in the room which became increasingly obvious as the Press conference went on. 'Speaking of animals,' said one wag, 'is it true that Sean Penn . . .' This hostility eventually moved George Harrison to observe, quite correctly, that most of us had predetermined our attitude.

'Are you,' he asked sweetly, 'capable of recognising the truth?' Animals had slipped into the general flow of wit and epigram when George suggested that sections of the Press had behaved like animals. This was greeted with a low growl from the floor.

'What do you think of England, Madonna?' 'It must be lovely somewhere,' she answered rather sadly. In response to a Scandinavian question, Madonna told us that there are no naked nuns in *Shanghai Surprise*. Apparently they are keen on naked nuns in their films in Scandinavia. Her next record, she replied to one of few questioners who wasn't either trying to establish his or her own credentials or wasn't apparently half-witted, is due for release in June. It will be called, I think, 'Live To Tell'.

My question, which was to have been to ask for Madonna's autograph for my children, went unasked. I was afraid people might think I was showing off.

# Madonna 2

## The Snoozing Beauty

*Observer*, 23 August 1987

---

'TWO-DIMENSIONAL' suggested the *Guardian*'s man at the Madonna concert at Wembley with the £5 programme and the Bacardi and lime. 'As good as that?' I muttered to my wife as we sat in the traffic after Thursday night's concert.

Before we take another step, I should emphasise that I like Madonna records; which, as they echo without respite from our William's bedroom and what is left of my radio-cassette player, is just as well.

We arrived at our seats to see the Bhundu Boys trooping from the stage. We may well have thus missed the most invigorating part of the night's entertainment. The Bhundus were replaced by a rather truculent young man bent upon persuading the music lovers at the front to move back a step or two. He might as well have asked them to set Roget's Thesaurus to music. 'Please take it easy,' he urged mournfully. Do people traipse to Wembley under any circumstance bent upon taking it easy? I think not.

Outside, the Gents' toilet was crammed with women who had wearied of queuing for the Ladies. I admired their initiative, which seemed seriously to embarrass male comfort-seekers. 'She'll start at 8.18 and end at two minutes to ten,' one of the women said.

At 8.18 Madonna appeared, wearing the fortified liberty bodice that has thrilled millions world wide. 'Thank you and hello, London,' she improvised. 'England,' she added helpfully. 'Are you gonna make this a night to remember for me?' she asked. With four children at my side and a £64 hole in my bank balance I murmured something to the effect that we were paying her to make this a night to remember for us.

From the first number, 'Open Your Heart', it was clear that Madonna was resolved upon obscuring what is, after all, a rather characterless voice by playing a sort of adjusted principal boy role – lots of costume changing and vigorous movement interspersed with the occasional 'shit' to remind us of her street origins or mention of her 'tits' or 'knickers' in case we forgot that Naughty Madonna is Sexsational.

The women in front of me, six strong and in their early twenties, were debating as early as 'Papa Don't Preach', the fourth song, whether to stand up and dance. Four more songs had passed before three of them did. As Madonna, playing 'Sleeping Beauty', feigned sleep on stage, torpor was stealing over the audience. All around Wembley anxious residents were fighting to stay awake.

'This is "Causing a Commotion" – and that's what I'm gonna do,' claimed the expensive song thrush incorrectly. This was a top class supper club set translated to stadium status and losing in translation.

'Is it the tit business again?' she asked hopefully when a section of the crowd sang to her. Apparently it was. Madonna is not, one imagines, troubled by self-doubt. She and her hugely gifted handlers have worked to make much of a modest talent. At Wembley I suspect I was not alone in being disappointed by her performance.

# Medieval Medicine

*Radio Times*, 5–11 November 1994

YOU WOULDN'T THINK it to look at me now, a back as broad and as fair as the Carpathians in spring, legs like mighty pillars of oak, but I was a sickly child. My school years passed in a miasma of vapour rubs, inhalants and other reeking nostrums and none of my tiny body's orifices was altogether safe from matron's probing fingers and her range of cures so dubious that it is possible that the drugs companies no longer peddle them even in the Third World.

My mother had a somewhat morbid interest in plagues and plagues-people and had, at some stage, acquired a book on medieval medicine. I remember glancing at this at an early age and being strangely moved. We still have it somewhere although I can't for the life of me find it. Given the prevailing enthusiasm for quackery, I have often considered opening a medieval medicine clinic somewhere fashionable in London. I feel certain that business would be brisk.

One of my favourite cures was for something called elfshot. The symptoms of elfshot were a general lassitude and grumpiness, a feeling that the best start to the day was the distant tolling of the bells rather than bird song or the Steve Wright show. The cause of elfshot, as you may have deduced, lay in the antisocial behaviour of elves who, having gained entry to married quarters, would fire tiny arrows into your sleeping body. I don't know about you, but I often have a touch of elfshot in the morning.

One of the cures – don't try this at home without checking with your doctor first, by the way, especially as I may have got it slightly wrong – is to cut yourself a length of birch, strip the bark from it and throw it between your legs into running water. Hey presto! No more elfshot. Another remedy, I forgot for what, involved pressing neighbours into service to coat your naked body with duck fat before flinging you into a hedge. I can visualise the arts-pages set falling for this sort of thing bigtime.

# Kylie Minogue

*Observer*, 10 July 1988

---

'ANYTHING STICK out like a sore thumb as being great?' Bruno Brookes wanted to know. Kylie Minogue looked puzzled – as well she might. The young Australian, one of the stars of *Neighbours* and further celebrated for the powerplays 'I Should Be So Lucky' and 'Got To Be Certain', was in the Radio 1 studios being interviewed by the influential DJ.

Five minutes earlier Kylie had stood, a disconcertingly slight figure signing autographs in the reception area, whilst I hovered, inconspicuously, hoping to catch some unguarded aside or witness some indiscretion that I might pass on to you. Timing my run to perfection, I contrived to enter the Radio 1 lift inches from the chanteuse and even pressed the button that took us to the second floor. 'Kylie, this is John Peel,' someone said. I dimpled prettily, Kylie smiled the smile of someone who has just flown in from Hong Kong and is going to spend the next few days being introduced to people to whom she would quite likely wish to take a chainsaw.

Outside the studio, she was greeted by Dave Lee Travis, an otherwise likeable man capable of reducing me to gibbering rage by referring to the BBC World Service, as he invariably does, as 'the BBC Wild Service'.

'It's Kylie Minogue,' he cried, deliberately mispronouncing her name Killie Minnogyou. Instead of headbutting him, she summoned, from the very depths of her being, another wan smile. On air, Bruno was offering listeners the original, rejected artwork for the LP by Voice of the Beehive. (It was won by an Andrew Neill.) Kylie, dressed in an off-the-shoulder scrap in red, went in to join him.

Under interrogation, the thrush revealed that she was 'not sure' whether she would return to *Neighbours*, confirmed that the shoes she wore for the photograph on the sleeve of one of the singles – 'What size feet do you take?' asked Bruno – were 'just a bit of fun', hinted that the United States is 'a whole different world', agreed that *Neighbours*, too, is 'just a lot of fun' and leaked that someone called Daphne is headed for a whole heap of trouble.

Bros, Kylie volunteered when pressed, were 'all right' and, responding to Bruno's 'Many people are going to Australia now', gave us a state by state guide to that country's meteorology.

Half an hour later we were both in the Salle du Jardin of Maxim's de Paris (actually Maxim's de Panton Street, SW1), having travelled thither under separate cover. The Salle was dotted with cardboard cutouts of Kylie wearing an expression so manic as to excite the attention of alienists. Kylie herself was being photographed a lot.

A further quarter of an hour later, Kylie was disappeared so that she could be reintroduced to us by Pete Waterman of wildly successful pop producers Stock, Aitkin, Waterman, who, after swearing at us by way of greeting, presented his star with a gold disc for what was called 'I Should Be So Bleeding Lucky'. This simple but strangely affecting ceremony was followed by a photo opportunity, after which Kylie made a brief and quite inaudible speech before being redisappeared.

# Misty in Roots

## Misty in Roots and 3 Mustaphas 3 in Cambridge

*Observer*, 15 June 1986

---

'WHEN WE TROD this land we walk for one reason. The reason is to try to help another man to think for himself. The music of our hearts is roots music, music which recalls history, because without the knowledge of the history you cannot determine your destiny.'

So runs the introduction to *Live at the Counter-Eurovision 79*, the first LP issued by Misty in Roots and still, seven years on, a record of the most remarkable musical and spiritual force. Misty have been treading this land and many others for twelve years now. They came together to back the touring Jamaican artist Nicky Thomas in 1974, subsequently establishing, in Southall, the People Unite Musicians' Cooperative and, in 1978, an arts workshop which, as it attracted a substantial number of young blacks, also attracted the attentions of the constabulary. During the Southall riots of 1979 a police attack on the workshop resulted in the destruction of thousands of pounds-worth of equipment and the hospitalisation, with severe head injuries, of Misty's manager, Clarence Baker.

Misty have sought to combine the spiritual traditions of Africa with the rhythms of Jamaica, whilst deriving inspiration from their experiences growing up in England. From early work with Rock Against Racism, the musicians have taken their message around the world, from Zambia and Zimbabwe, which inspired the LP *Musi-O-Tunya*, to Poland and East Germany. Last year they became the first reggae band to play in Russia, winning over what appeared to be a largely hand-picked audience of life-in-the-slow-lane functionaries.

On Thursday Misty played King's College, Cambridge, to an audience which must have struck them as being quintessentially Babylonian. The vicious acoustic of the vaulted, stained-glass-windowed dining hall had earlier played what wordsmiths described as havoc with the Balkan keening of the 3 Mustaphas 3. The Mustaphas, since touring recently as the kernel of a big band, have developed a more fluid line-up and you, the eager consumer, can, I was told in halting English by a colourfully dressed member of the ensemble, pretty much have the Mustaphas of your choice these days.

When Misty played, the disorderly acoustic seemed suddenly tamed, the dense, rolling sound of the nine-piece recapturing the power of the 1979 recording. But in 1986 there is a spring to Misty's step, a greater emphasis on melody and light and shade, changes highlighted by the presence on stage of a young, snappily dressed saxophonist, *très, très Face*, alongside the three-man vocal front line.

Standing in the gallery overlooking the bobbing heads of the predominantly white, predominantly middle-class audience, and listening to the sweet, sad voices of the trio, men who look like Old Testament prophets after a particularly fraught month or two on locusts and brackish water in the wilderness, in the spectacular if fussy King's setting, was likely, on the face of it, to prove a rum experience. Yet Misty contrived, as they must have in Warsaw and Moscow, to absorb setting and audience into their work and world, the incongruities melting away to nothingness.

The frankly anthemic 'Own Them, Control Them', soon to be a single, was inspiring. Below me a reveller apologetically removed his boater and danced. Perhaps he was the braying ninny who had catcalled after me in the street outside, demonstrating what he and his chortling cohorts clearly felt was wit of Johnsonian weight. Misty in Roots make you believe that anything is possible.

# Montel

*Radio Times*, 2–8 December 1995

SHE 'GAVE HIS SUITS to a Neighbour'. He 'Set Fire to Her Parents'. As a strictly sound-off televiewer, at least twixt sun-up and sundown, I see loads of popular confessional chat shows but hear none. I am more likely to catch Ricki Lake or Oprah than our own Vanessa Feltz. As for Esther, well, years ago I decided that I would rather be broken on the wheel than watch anything featuring her – although I was tempted recently by the programme billed as 'Rantzen and her audience discuss music with DJ Tony Blackburn'. What a clash of titans that must have been.

Much has been written over the years by people amazed that so many of their fellow human beings are prepared to reveal their innermost thoughts and secrets to a global television audience while wearing unsuitable sportswear. I have secrets too, of course. I once drank a pint of bitter in Shrewsbury when I was only fifteen, but I don't think 'He Drank Pint of Bitter When Fifteen' is quite enough to capture a big audience.

More promisingly, I urinated on a chair that I knew would shortly be occupied by someone with whom I had but recently had a bit of a disagreement. (We're talking soft furnishings here, you understand.) But I'm not about to sit, powdered and sleek, next to my victim and in front of the cameras and admit to my moment of shame.

I know I'm no one to talk, but one thing I have noticed about a clear majority of the television confessors is that they tend to be rather extraordinary shapes. Perhaps it is the camera that makes men who have 'Spurned Offer of Love from Mother-in-Law' and women who have 'Given Up All to Study Ukulele' seem to be triangular, but I think not.

(I am writing this as I watch, with the sound off, of course, something with a title along the lines of 'Truck Trail Europe 95'. This involves a lot of men who look like Hale or Pace. It also involves, less predictably, women's bottoms. I have nothing against bottoms – in fact, I have one myself – but I question their use in motor sport. Here is Rudolf Reicher Jr in his Zil. I don't think Rudolf is going to show us his bottom though.)

Where was I? This week another programme enters the chat show/intimate confessions scrum. Channel 4's *The Montel Williams Show* is introduced by Mo . . . Oh, you're ahead of me on this one. I like to think that Montel was christened Montelimar but abbreviated his name to avoid being the butt of jokes at school.

Montel is from Baltimore and was, it seems, the first black enlisted marine selected to attend the Naval Academy Prep School. Later he studied Russian and Mandarin Chinese, along with general engineering, international security affairs and cryptology. There must be a joke in there somewhere but I can't be bothered to winkle it out. Perhaps you can.

Montel, the American press release continues breathlessly, 'has travelled extensively'. In the cruel, brittle world of publicity, anyone who forgets to follow the word 'travelled' with 'extensively' is made to stand in the corner.

Montel has, in addition to travelling extensively, done loads of marvellous things. He has won 'numerous awards and distinctions'. He also has a skipload of medals, including the Armed Forces Expeditionary Medal, the Humanitarian Service Medal and two Meritorious Service Awards. He has done three years of motivational speaking. He has been recognised ('Hey, isn't that Montel over there?') by the USA Chamber of Commerce with a Special Services Award, joining Ron 'n' Nancy Reagan and George 'n' Barbara Bush. He has been featured 'on every national news program [sic] in the', er, 'nation'. Miraculously and presumably as the result of bold initiatives being taken in the field of cloning, Montel has also 'travelled the country full-time'.

And what is this Mandarin- and Russian-speaking, youngster-motivating, engineering Pentagon paragon (good that, isn't it?) featuring in his first programme? Why, interracial wife-swapping. On ice? Not this time, I'm afraid. But stand by for synchronised interracial wife-swapping in the next Olympics but one.

If you plan to watch *Esther* this week, by the way, she has 'Making a Naked Living' on Monday, 'The Adam Faith Interview' on Tuesday and 'Mixed Race Relationships' on Friday. I'm afraid you are going to have to craft your own joke here as well.

# The Mourning of the Golden Flask

*Sounds*, 19 April 1975

TIMMY BANNOCKBURN, bless his little cotton socks, has broken something called the Golden Flask. Each morning about this time he teaches us all a nursery rhyme, reading out the age-old world like Christopher Lee on downers, and urges us, with many a boy-next-door chortle, to put the kettle on, make a nice cup of tea, and settle down for a coffee-break.

This is, for me, the high-point of his programme, for having played us a nice Jack Jones record or a nice Shirley Bassey record, he takes up a moment or two of his time to talk to us about something close to his heart.

Last week – as he does most weeks – he spoke to us about critics, most particularly those critics who find his programme infantile, insulting and just plain old wet. These worthless hacks (my words, not Timmy's) are (his words, not mine) either frustrated artists themselves or they just have a chip on their shoulders. So the implication is that Tim considers himself an artist. Now there's something to think about.

However, this morning we heard nothing about critics, about Simon (his little boy), or about the trade unions. Instead, after a brief homily on the kindness and generosity of the people of the

West Country, he told us about the breaking of the Golden Flask.

At the time I was edging my Renault 4 through the rain-soaked streets of Tooting, past a very ordinary little terraced house which carries a plaque claiming, with a sort of mute defiance, that Thomas Hardy once lived and worked there, and was thus unable to make myself a cup of tea and join in with the coffee-break festivities. However, I mourned, along with millions of others, the passing of the Golden Flask. Timmy was clearly concerned about its passing, he spoke for quite some time on the matter, and I and my passengers, still slightly elated after the old Jackie Trent record he'd just played, grew silent. Our brows were, I confess it, furrowed – and furrowed deeply!

After several minutes devoted to a eulogy to the departed Golden Flask, Timmy played us another record – I think it was 'Ding-A-Dong' although Tim called it 'Ding Dang Dong' (he's so playful!) – then returned to the topic of the lost-and-gone-forever tea flask. He was wondering out loud, allowing us to share in his innermost thought patterns, how he would find a replacement for the dear departed. In fact, he spoke about it with such feeling that we got the impression that he was dropping the tiniest hint that we might just go right ahead and send him a brand spanking new Golden Flask. And who says that British radio is unlistenable, eh? Why, it is crammed to overflowing with the very stuff of life itself.

The Pig and I are resolved to go out tomorrow and buy Tim a Golden Flask – and we were just saying how nice it would be if all the readers were to send the man a message of sympathy at his most distressing loss.

The high-point of last week – aside from Bannockburn's thoughts on critics – was a visit to Ronnie Scott's to sit and watch Arthur Brown and his new band puzzle the show-biz elite. I missed about half the show – I was on my way thither after finishing one of my wunnerful radio programmes – but arrived to see Arthur, the grand old loony of British music, prancing and dancing with all the vigour he showed years ago during the Acid Summer of 1967. As always he was entirely unpredictable, singing at times like a zomboid Tom Jones, then like a hot gospeller in the throes of getting right with de lawd. In fact, Arthur was accompanied at one point in the rather bizarre proceedings by a quartet he introduced as the Gospelaires. Even they, presumably accustomed to the dervish-like ecstasies of those filled with the spirit, looked somewhat confused by the wild cavortings of the man Brown.

He closed his show – for no reason that I have been able to fathom – by bringing forward a detachment of Morris dancers, who skipped and jangled rather awkwardly in the narrow confines of the club. Arthur has brought us a lot of interest, a lot of amusement and a lot of good music down the years, first with the Crazy World of, then with the admirable Kingdom Come, and it should gladden your hearts that once again he has elected to pass among us. There is, indeed, a new LP available on Gull called *Dance With Arthur Brown*, which I hope one of the bright young sparks on this journal will have reviewed or will review.

I'd like to stay with you longer, but someone has just brought me a record of the song that came third in the recent Eurovision excitement. In case you've forgotten, this was – and is – 'Fallin'' by Wess and Dori Ghezzi. I must away to hear it.

# Napalm Death

*Observer*, 4 October 1987

---

FROM THE PRESS enclosure, an upturned beer crate behind the bar, all it was possible to see over the crowd – the temptation to write 'sea of humanity' here is very strong – was a strip of light 30 foot across and a foot deep, the space above and beyond the heads and below the ceiling. Napalm Death were playing, but only their bassist was visible. Of the singer, barking and yelping in a manner that would have given H. P. Lovecraft sleepless nights, there was no sign. Napalm Death, whose LP *Scum* boasts twenty-eight tracks of quite exceptionally rapid music, described on the sleeve as 'savagely brutal hardcore thrash', sounded as though they were playing soundtracks of the end of civilisation. I liked them a lot.

The heat inside Nottingham's Garage was enough to have caused humming birds discomfort, so I accepted an invitation to observe the rest of the night's entertainment from behind the stage. From a position on a flight of stone steps running with condensation, I watched Heresy, hymned in these pages several weeks ago. During a performance impaired by sound problems, several hundred of the nation's young, principally male and evidently oblivious to pain, expressed their appreciation of the uproar by seemingly attempting to fuse their bodies into one hellish unit by main pressure. Although they failed – but gloriously – to achieve this, it was generally impossible, when one caught sight of a flailing arm or leg, to ascertain to which body it belonged.

On stage, several formidably constructed young men were attempting, good-naturedly, to prevent life-forms separated from this heaving entity from falling forward and becoming as one with the amplification equipment. When Heresy yielded to the American band Holy Terror – bigger, healthier youths in shorts and stylistically nearer to heavy metal – there was a perceptible rush of denim to the front and some unabashed head-banging. The be-denimed ones, who, judging from my research, came principally from Chesterfield, were restrained when compared with the hardcore set and there was none of that lunatic hurling of bodies on to the stage, somewhat in the manner of American footballers seeking a touch-down, witnessed earlier.

'You guys are crazy,' suggested Kurt, singer with headliners DRI (Dirty Rotten Imbeciles), genuinely taken aback that in this preposterously overheated cellar some of the gibbering creatures before him were wearing jackets. By now the area available for performance had been reduced to the point at which Kurt's bottom rested against the drum kit and his hair threatened to intertwine with that of the spectators. During one song he crawled out over the heads and just below the dripping, black ceiling in a manner reminiscent of a miner, eventually rolling back to the stage on his side. Finally, as the band raced insanely about him, Kurt curled up on the floor and howled his songs from a foetal position. This had been another elemental night in Nottingham.

# New Age Music

*Observer*, 7 February 1988

THERE CAN BE few areas of musical activity where the description 'perfect dinner party music' would be seen as anything other than a call to arms. Yet Coda Records, the right martyrs of British New Age, like this assessment of the work of their Dashiell Rae well enough to reproduce it in publicity material, an attitude sufficiently weird to draw me to the Purcell Room for the second of three nights of New Age music.

The Coda publicity crew seemed to be of the same school as those who write menus for Indian restaurants, rattling out such stuff as 'music to relax but still occupy the mind', for example, or 'mind-stretching – yet in part ethereal'. Then there is '. . . that special relationship between the earth and sky. Most pronounced at dawn and dusk, the union of the two can sometimes be seen but never touched.'

This last is associated with the piano work of Eddie Hardin. Eddie, whose 1969 LP *Tomorrow's Today* boasted such quintessentially New Age titles as 'Candlelight', 'Beautiful Day' and 'Mountains Of Sand', opened Thursday night's proceedings. As Eddie played, I occupied my mind with attempting to discover common characteristics within the audience. Apart from pink cheeks and an interesting tendency towards heads blessed with one or two bumps over and above the standard issue, I could find none.

No less an authority than Trevor Dann has told us that New Age music is 'serious', but it was difficult to take this wafer-thin romanticism seriously. After forty-five minutes, several customers slept deeply, but this, reason insists, would surely please the New Age performer, who would presumably regard a standing ovation as proof that the beastliness had not been stilled. A twenty-minute interval allowed us to recover ourselves before plunging back to watch John Themis. It is he whose music is claimed to 'relax but still occupy the mind'.

Assisted by keyboard operative Terry Disley, Themis, possessed of a dazzling guitar technique, approached his work with unexpected levity, affecting a jovial manner and a black beret. Supported by a cheering section of rough-hewn men whose own manners spoke more of the Marseilles waterfront than of the pine forest, frozen lakes and waterfalls beloved of New Age art departments, Themis and Disley demonstrated how electronics can combine to make anything sound like, well, anything – at a single touch your new keyboard can sound like a sofa.

As the set induced passivity and I became convinced that I could feel my brain cells dying, I wondered whether New Ages finished with an audience participation number. Would we, I mused, as sleep tiptoed up behind me, be invited to hum along? Woken by applause some minutes later, I scribbled 'blithering nonsense, superbly played' and bolted before the threatened encore.

# New Year's Eve

*Radio Times*, 19–25 January 2002

---

WILLIAM WAS ON the phone from Newcastle. He wanted to know about avocados. 'What aspect of avocados?' I queried. 'Their role in the War of the Spanish Succession, perhaps?' William wasn't listening. 'Is Mum there?' he asked. Our children know that their mother does the handy home hints and that I do the feeble sarcasm. It was the afternoon of New Year's Eve and panic was in the air. The problem was that the media combine to persuade us all, particularly the young, that the ultimate good time is out there waiting to be had on New Year's Eve and that somehow we are diminished as people if we fail to seize the opportunity to party, party, party. William was preparing dinner for friends, something his father would be incapable of doing. I have no idea what to do with an avocado other than to eat it. I could throw it up in the air and catch it, I suppose, but that's not enough really, is it?

Tom, William's brother, was driving to Nottingham in search of his good time. Attempting to discuss his plans with him as he hurled about the house was frustrating, but I got the impression that he was lukewarm about Nottingham. His friends were putting on their best suits and going to a chic bar. 'In *Nottingham?*' I queried, but Tom was gone. His last words before he left were to the effect that he might go to Sheffield instead.

Flossie spent a lot of the afternoon on the telephone. 'I'm not telling lies,' I heard her insisting at one stage, and with my instincts for self-preservation on red alert, I decided against asking what was going on.

We knew where Alexandra was going to be. She was joining her mother and me in front of the television here at home. As Alexandra had waited on tables in a rather agreeable pub in a small village on the outskirts of Ipswich in the run-up to Christmas, you might assume that she would have made a small fortune in tips from kindly revellers and therefore be in a position to experience a New Year's Eve of almost Babylonian debauchery. You would be quite wrong. One night, for example, she waited on a table for twenty-five who ran up a bill in excess of £700 and left Alexandra not a penny.

We were spending the night at home for several reasons, the principal one being that our friend Laurie died over Christmas and we didn't feel in the party mood. Laurie was one of a small number of people I have known that I have admired unreservedly. Among the happiest days of my life were those in the eighties when Laurie and I cycled out across East Anglia almost every day with a pub as our short-term destination and extreme fitness as our unrealistic goal. Laurie Self, fifteen years my senior, was as gentle and funny a person as I have known, and if there is any good in me then at least a part of it will have come directly from my friend. Four of Laurie's paintings, a distillation of the spirit of the Suffolk countryside through which we cycled together, hang on our walls, but we won't need them to remind ourselves that here was a lovely man.

# 1975
## Searching for Stars

*The Listener*, 25 December 1975 and 1 January 1976

---

IN HIS REITH LECTURE, 'The Power to Leap', Daniel J. Boorstin spoke briefly of that twentieth-century curiosity, the celebrity 'well-known for his well-knownness'. In 1975, funsters of this particular breed have dominated rock – a serried rank of glittering monsters created, according to best scientific principles, by adept promoters, and given life through the morbid craving of the rock audience for a Messiah a year. This laboratory exercise has left rock music with a dangerously unbalanced social order, with a small and preposterously wealthy upper class, no middle class, and a vast working class. The gulf has become so great that the observer may be reminded of various sun-soaked, guerrilla-infested, Latin-American republics, where a tiny minority controls most of the wealth and the rest do the best they can with what remains, scornful of the ruling elite and fervently wishing to join it.

The rock star well-known for his well-knownness may seem to be an enviable creature, but he survives by his artistic sterility and is enslaved by his success. His status has become such that he (or she, or they) need perform only three or four times a year in some vast sports stadium or other, usually, we are assured, at a considerable financial loss, and thus as a favour for his fans, who, sadly, are prepared to put up with all this nonsense. Their records must reflect those ingredients that have made previous works successful; to develop artistically is to risk oblivion. The inevitable result is an awful blandness and, indeed, blandness has become the key to continued success, particularly in the USA.

At the end of 1974, the top ten LPs in the States were by Elton John – who was at number one with a selection of his greatest hits – John Denver, Harry Chapin, Neil Diamond, Bachman-Turner Overdrive, Loggins and Messina, Ringo Starr, the Rolling Stones and Jethro Tull. I know that is only nine names, but John Denver was in there twice. At the end of this year, excitement, as represented by the Stones and, maybe, Jethro Tull, has disappeared completely, and the top ten are Chicago, who are at number one with a selection of their greatest hits, Jefferson Starship, Elton John (twice), John Denver, the wonderfully colourless America, Seals and Crofts, who make even America seem thrilling, The Eagles, Paul Simon, Art Garfunkel and Joni Mitchell.

Nothing there, apart from Joni Mitchell, to make you stiffen the sinews, summon up the blood. In fact, for the first time in years, the British LP charts hold more of real interest than the American, with the Pink Floyd, John Lennon, Supertramp, the Rolling Stones, Roxy Music, Mike Oldfield, Queen, Steeleye Span and Rod Stewart up there, alongside the Bay City Rollers and James Last.

Fortunately, for those dismayed or just plain bored by the star machinery and its wholesome and non-nutritive product, there is still a wide range of alternatives. At the start of the year, it seemed as though black music, through the medium of the discothèque, was likely to make serious inroads into white and coffee-coloured domination. Sadly, the machine, sensitive to the dangers inherent in any lively alternative, has all but flattened the disco scene in less than a year, and many of the currently acclaimed dance records come from session musicians in studios in France and Germany. Reggae, the music of Jamaica, found chart respectability in 1975 in the person of Bob Marley. The extraordinary business practices that govern the recording and distribution of reggae, and the complex social/political/religious background to the best of the music, may make it last longer than most fads, and make it particularly tricky for the big-time promoter to take over. Country music, too, has built on its traditional strengths in rural Britain, with Tammy Wynette reaching the top spot in the singles charts during the summer and Waylon Jennings reaching the ear of the cultists. Then there is African music, which has advanced far enough to have its own chart, printed at irregular intervals in one of the top rock weeklies.

But if the Jamaican Marley is set fair to make the quantum jump into the premier league, are there any British musicians poised to do likewise? Queen, described as 'Britain's latest show-business rage' in that noted rockaboogie guide, the *Daily Express*, are already in the air, and I still believe that the superb Be-Bop Deluxe and the scarcely less worthy Thin Lizzy will follow. I must confess that, given the artistic wasteland that surrounds large-scale success, I almost hope they don't make it. If you like to see it as some sort of belated revenge for the American War of Independence, you can rejoice that we seem to have fooled the Yanks into buying Bay City Rollers, too. In return, we have been offered two American Messiahs, have partially accepted one and will, in all probability, accept the other.

The one is Bruce Springsteen, who was hailed by the usually responsible *Rolling Stone* as 'the future of rock 'n' roll'. In truth, Springsteen offers us an enjoyable supper-club pastiche of rock's brief history, served up in *West Side Story*-styled tat. The other is Patti Smith, poetess. Two of her public statements should give you a fair guide to the woman, in addition to qualifying her for membership in perpetuity of 'Pseuds Corner'. 1. 'Rimbaud would have made a great lead guitarist.' 2. 'I didn't like white music. But then Jagger came along. All of a sudden I knew what to do . . . drop my pants.'

Bruce and Patti are, then, the new generation of well-knowns, arriving on the scene at a time when the tomfoolery of rock stars has moved from its rightful place in the *Melody Maker*'s 'Raver's Column' to William Hickey's section of the *Daily Express*. What, I expect you are wanting to know, have the old guard done to enrich our lives in '75? Well, Elvis put on a lot of weight and Elton John became, on 7 June, the first artiste to have a new LP go straight to the top of the American charts in its first week of release. This was *Captain Fantastic and the Brown Dirt Cowboy*. Bob Dylan went on a low-key tour of (roughly) New England with a battery of chums that included Joni Mitchell, Joan Baez and Mick Ronson. This was a good thing. David Bowie stayed in America and away from our taxman. So did Rod Stewart. Both made good LPs. Rick Wakeman performed his splendidly silly *King Arthur* piece, complete with a cast of tens, on ice at Wembley. The Pink Floyd followed the classic *Dark Side of the Moon* with *Wish You Were Here*, and all the reviewers agreed it was very poor. Mike Oldfield followed up the follow-up to *Tubular Bells* with *Ommadawn*, and all the reviewers agreed it was very good. Harvey Smith made a single. So did John Conteh. One or two bands, mainly from the second and third divisions, broke up. They included Humble Pie, Lindisfarne, Traffic, the Groundhogs, Brinsley Schwarz, the New York Dolls and Chilli Willi and the Red Hot Peppers.

Much of what I have written above may seem to be either unduly cynical or pessimistic. In particular, my remarks about the damage done to the intellect of the creative artist by the lotus-eater syndrome of success, and by the inevitably uncritical adoration of his or her fans. But there is still hope that many of the performers thus afflicted can and will produce work, if not in 1976 then, perhaps, in 1977, to match the very best of what they have done. I say this because of two remarkable examples this year of an individual's ability to recapture whatever vision it was that fired them at the start of their careers. The first example is Joni Mitchell; the second, Neil Young.

I have never been a great admirer of Joni Mitchell's work, feeling that the considerable promise of her earliest work had disintegrated into a James Taylor-like welter of self-pity and self-indulgence. However, her recently released LP, *The Hissing of Summer Lawns,* is magnificent. The performance of Neil Young is, if possible, even more remarkable. Since his classic *Everybody Knows This Is Nowhere* LP of 1969 Neil has shaped up like the classic show-business casualty, with a series of depressingly poor releases, rumours of major problems with narcotics, and few and indifferent public appearances. Although it may well be that Neil Young's personal problems are far from over, he has contrived to harness them to the production of a masterly LP, *Zuma.*

I take heart also from the appearance of Laurel and Hardy in our charts, from the continuing excellence of the releases from Island and Virgin Records, and from the spirited revival of EMI's once floundering Harvest label. This latter has brought us two magnificent albums from Japan's Sadistic Mika Band in 1975, and one apiece from Roy Harper and Be-Bop Deluxe. And *I* did like the Floyd's *Wish You Were Here,* which was also on Harvest. In fact, *Wish You Were Here* must be my own Record Of The Year, closely pursued by *Zuma, Hissing of Summer Lawns,* Little Feat's *Last Record Album,* Dylan's *Blood on the Tracks* and Tangerine Dream's *Ricochet.*

Best new group name of the year? Duke Duke and the Dukes. Hope for 1976? That Little Feat will come to Britain again.

# 1977
## Banned Bands

*The Listener*, 22–29 December 1977

---

AT THIS TIME last year, a snap of the Sex Pistols' Johnny Rotten appeared in these pages, and it would be foolishness to pretend that this same Rotten has not, both as singer and as emblem, been the focal point for the British rock audience in 1977. The Sex Pistols, despite the tendency of their manager to operate them as a freak show rather than a band, have contributed three classic singles and a matchless album to the year, with one of these, the single, 'God Save The Queen', demonstrating, in Jubilee year, many of the sillier inconsistencies in our national character.

Because of the Sex Pistols and the remarkable number of powerful new bands to spring out of nowhere in their wake, this has been the most rewarding year for the rock fan since 1967, perhaps 1964, possibly even 1954. Mind you, it would be asinine to claim that all of punk/new wave has been good, even though a surprising amount of it has been. This autumn, we have witnessed a certain indecision in the punk ranks as the first exhilarating headlong rush has ended. The defiant self-confidence and the exuberant bitchiness of the spring have evaporated, leaving in their stead bitterness, recrimination, and a growing suspicion that the leading punk bands aspire, despite their early protestations to

1972: John at home with Eddie Cochran

**JOHN PEEL**

*... is not in this bit. We had deadline problems and so did he — unfortunately they didn't coincide. Normal service will be resumed next week.*

*Sounds*, 29 March 1975

*JOHN PEEL promised to resume his column this week, but some football match or other got in the way and he was suddenly taken drunk, as you can see from this picture. The merciless lens of Ray Stevenson caught him in drunken revelry at the Speakeasy, where he was prevailed upon to introduce the farewell appearance of the **Snivelling Shits** (above), who are off to Paris to inflict some unhappiness on the Frogs.*

*Sounds*, 20 May 1978

**PEEL**

**We regret that we have had no communication from John Peel since he last reported sunbathing at a campsite near Como. Consequently there will be no Peel column this week. Anyone finding a message from Peel should send it to Disc where postage will be refunded.**

*Disc*, 21 August 1971

**Above:** 1969: John and Sheila on their first holiday together near Athlone, Ireland, with Adrian Henri and Andy Roberts

**Below:** 1971: John not playing mandolin with the Faces on *Top of the Pops*

**Above:** At home, sorting through demos in the dining room

**Top left:** 1985: Record shopping in Bremen, Germany

**Below:** In the studio at Radio 1 with producer John Walters

leads to a ferocious assault on the form, a process which, in this case, led to those bodies which control radio and television advertising banning advertisements for the Sex Pistols' subsequent LP, observing that their objection lay not in the style or content of these advertisements, but 'it is the nature of the record itself we object to'. A dangerous precedent has thus been established.

As several others have done before them, the Sex Pistols contrived to appear in that invaluable shop-window, *Top of the Pops*, while convincing the gullible that they were reluctant to do so, and were placed fifth in the annual poll conducted by the conservative *Melody Maker* behind those firmly entrenched dinosaurs, Genesis, Yes, Led Zeppelin, and ELP (Emerson, Lake, Palmer). More importantly, the Pistols romped over all opposition in the recently published *New Musical Express* poll, with the older bands relegated to third, fourth, second and nowhere respectively.

ELP released their first LP in two and a half years, the monumental and monumentally tedious *Works Volume 1*, following it with a *Volume 2* – you guessed – the hurried and opportunist nature of which may indicate that E, L and P feel that their bubble is finally set fair to burst. If it were not for the existence of the triple LP set, *Consequences*, wrought by former members of 10CC, a set which does have the advantage of being well-played, the two *Works* albums would stand unopposed in the Pompous Nonsense category for 1977.

[illegible] ...tect the interests of [illegible] community of which they approved against the interests of those of which they disapproved, started banning anything suspected of being a punk band from appearing in buildings over which they had control. In Stirling, they were saved the bother when outraged students hounded the luckless Damned from the stage under a hail of missiles. The Damned went on to refuse to perform at the year's only outdoor punk festival, the débâcle at Chelmsford. The idea was average, the execution poor, and, as compère, I was grateful to escape undamaged.

As usual, the spring brought rumours of impending rock festivals a-flooding, with famous folk, some still alive, supposed to be appearing at such places as Longleat and Potters Bar. As usually happens, the only one that did happen happened at Reading, and even that was uncharacteristically colourless. I compèred this, too, and was cast down when the mud-coated fans failed to respond to the music of my own favourite new band, the Motors. Their début album on Virgin, called *1*, brings together the best of the old and new in a most felicitous manner, and will, I suspect, grow in importance with time, certainly with archivists, hopefully with fans.

Television-programmers, whatever brief they hold, seem to suffer from a form of tunnel-vision brought about by their conviction of their own infallibility. The *Old Grey Whistle Test* people seem to have persuaded themselves that 1977 hasn't happened, and have, instead, focused their cameras on the United States and 1976 and beyond. My own radio programmes have expanded from one hour each week-night to two, thus allowing producer John Walters and me to demonstrate more effectively the best of what has been going on around us. Traditionalists have bombarded us with hostile letters as a result, many suggesting that my high regard for the new music could only be the result of unenlightened self-interest, and the lining of my pockets with gold by shrewd entrepreneurs. Why, they have wished to know, haven't I continued to support the interests of the established groups? In vain, I have tried to persuade them that the chances of being corrupted by representatives of the tiny labels which have provided the bulk of the punk/new wave records is nil. In fact, they are more likely to come to me for a loan. But these correspondents have demonstrated that there are many rock fans who feel as threatened by punk as the devotees of Johnny Ray, Frankie Laine et al. felt in the fifties when rock 'n' roll was first made flesh.

Radio 1 – is this more self-interest? – has continued its gradual improvement, introducing (said he, treading carefully) younger and less garrulous presenters for a marginally more interesting selection of records. One down, and one to go. The commercial stations continue to function, with too few exceptions, in a stupor induced by excessive self-congratulation.

I don't propose to discuss American rock in 1977 at length. As viewers of the ossifying *Whistle Test* will vouchsafe, it has become tedious beyond belief. In the spring, Stevie Wonder's fine *Songs in the Key of Life* was usurped at the top of the US album charts by the soundtrack recording from *A Star Is Born,* featuring the unappealing Barbra Streisand; and this, in turn, surrendered the reins of office to Fleetwood Mac's tuneful but dull *Rumours*. This has been there for most of the rest of the year, surrendering even as I type to an LP by Linda Ronstadt, a young woman whose assets are a face of cloying sweetness and a voice of the type reviewers describe as 'achingly pure'. Her speciality is butchering defenceless Buddy Holly songs.

Punk will not, I fear, make many inroads into the cultural smugness of America – Columbia Records have already refused to issue the Clash's superb début album there because it is not well enough produced – although, perversely, their better bands, Television, Talking Heads, Devo, MX-80 Sound, Père Ubu and lamentably few others may survive longer than most of our own new groups. Before I grow too chauvinistic, I had best remind myself that the Sex Pistols' album was supplanted at number one in Britain by *The Best of Bread*.

Enough – more than enough – has already been written about the deaths of Bing Crosby and Elvis Presley. Their departures served to underline that we have reached the end of an era in popular music – at least, I hope we have. Actually, few of my own prayers for 1977 have been answered, although the year got off to a bright start when the American *poseuse*, Patti Smith, broke her neck (not too seriously). Not nearly enough broken-winded bands of ancients dissolved – their tenacity the result, in most cases, of accountancy rather than music. The occasional loony forecast that the Beatles would reassemble, although John Lennon, tragically, announced that he would not record again. Elton John retired from active service, and the Rolling Stones should have done.

It would take far too much space for me to list those things worthy of your attention in the New Year, but note in particular John Cooper Clarke, a witty and perceptive poet; Sham '69 and their charismatic leader (although he himself hates such a description), Jimmy Pursey; and the Slits, the latter a great name with the potential to become a great band.

I would also have offered an invaluable list of records from 1977 that should be avoided at all costs, but the *Sunday Times*, by recommending, surely with tongue in cheek, twenty of the year's most notable horrors, has beaten me to it.

I am not a dancing man, my only concession to the art being a barely perceptible movement of the knees, but I have done more of this in 1977 than in any other year – and I look forward to doing a lot more of it in 1978.

# Noorderslag

*Radio Times*, 22–28 January 2000

---

AS I MENTIONED a couple of weeks ago, I have been excited to discover that I am still growing. Having been a sturdy 5ft 9in since I served my Queen and Country in the fifties and expecting to have lost, through natural wastage, perhaps an inch over the intervening years, I was awestruck when I attended an insurance medical and was diagnosed as being an imposing 5ft 10in. With good fortune, I suggested to our family doctor Ian, I might yet make that all-important 6ft. Ian smiled indulgently and told me to hurry up and get dressed.

This newfound loftiness did me no good at all, I'm afraid, on a Radio 1 trip to Holland this week. I've always believed that the Dutch were, by and large, too tall anyway and it seems that in recent years they have added a certain feistiness to this blatant height. I wasn't aware of this development when I first woke in the City Hotel, Groningen, and found myself listening to Steve Lamacq on BBC World Service urging me to celebrate the rebirth of upfront British guitar music. Nor was I aware of it when I finally figured out how to work the room TV and watched in astonishment as men and women with uncommon physiques and little or no dress sense engaged in what my father would have styled farmyard activities.

I finally discovered the newer, taller, pushier Netherlands in a stuffy basement room called the Catacombe watching Cirith Gorgor – I think the names of venue and band should tell you all you need to know about either – where I could see nothing but the backs of swaying Nederlanders whose heads threatened to engage with the ducting in the ceiling yards above me. As the room filled, I became fearful that I might be trampled underfoot. These giants were, as giants historically usually have been, quite happy to barge you out of the way if you stood between them and the bar, the exit or the other giant or giantess of their choice. Our Radio 1 team – Anita, Lynn and I – representing Britain at Groningen's annual feast of pop fun, the Noorderslag, got mightily fed up with being shoved about, let me tell you.

The Noorderslag is part of an event called Eurosonic. Or is Eurosonic part of Noorderslag? Either way, the weekend had started well for us with the appearance in the prosaically named Grand Theatre of the Moldavian band Zdob Si Zdub. I have to admit that our interest in Zdob Si Zdub stemmed more from the fact that none of us had ever seen a Moldavian band before than from any expectation that they'd be good. In the event they were both Moldavian and terrific, incorporating into their guitarbass-drums rush curiously wrought traditional instruments probably called the kuk and the zhpren, thereby initiating a new genre I have styled campfire funk. Afterwards, the theatre started filling with giants and solitary men of my own age with droopy grey moustaches and a desire to talk about Deep Purple, so we made our excuses and walked the few protes (the Moldavian measurement equalling 14.7 metres) to Vera, a venue in which Britain's representatives at Eurosonic, Hefner, were shortly to play.

Anita, Lynn and I already know how peachy-keen Hefner can be, but on Friday night they exceeded our expectations, ripping Vera apart. Their triumph was, I felt, at least the equivalent of a 0–3 away win, yet more impressive because the set started to cries from hecklers of 'We want Dutch bands', a curious thing to be shouting at what was billed as an international festival, especially as the hecklers heckled in perfect and unaccented English.

In the morning and emboldened by Hefner's triumph, we decided to seek out two record shops recommended by friends from Amsterdam. These were located in what, according to their scribbled notes, was Steen Straat. With the practised eye of the former soldier, I quickly located Steen Straat on the map and, although it seemed some distance from the centre of town, I encouraged my companions by suggesting it might be in some rather amusing student quarter in which they could replenish their stock of designer tops and we could take coffee and fattening pastries in a café filled with ganja smoke.

Thus heartened, we set off for Steen Straat. An hour later we stood on an extremely run-down estate, on which, judging from the nature of the mounds of excrement piled on the pavement, large wild animals roamed free. There were no record shops on Steen Straat, only boarded up hardware stores, a butcher's and a hairdresser's. I suggested to Anita and Lynn that we could allay suspicion by having our hair done, but they weren't keen. It started to rain.

An hour later, wet and with aching limbs, we were back at the City Hotel. I had another look at the map and at the addresses I had been given, discovering that the record shops were on Steent Straat rather than Steen Straat and that the hotel backed on to Steent Straat. The shops on Steent Straat were just shutting. I was not popular. Silently we walked on aching feet over to the Noorderslag for several hours of being shoved about. I'm not going next year unless I have passed that 6ft mark, that's for certain.

# Oddballs

*Radio Times*, 9–15 July 1994

---

THERE ARE CERTAIN key words in the lexicon of promotion that lead to the suspicion that the human spirit is about to be dragged away by a cackling zombie and walled up in a dank cellar filled with rats. Principal among these is, of course, 'zany'. ITV's *Oddballs* is not described by Carlton as 'zany', but the words 'goofs', 'hilarious', 'cringe' and 'top sporting celebrities' do occur in their press release. In fact, 'top sporting celebrities' occurs twice.

*Oddballs* will bring Eamonn Holmes into our homes and he will be introducing what are described as 'classic sporting funnies'. Eamonn is himself described as a 'self-confessed sports fanatic', which seems to imply that even better-known presenters employ people to do their confessing for them.

We will also, it seems, be invited to chuckle at 'some of the more bizarre sporting events to be found in far-flung corners of the world'. This is promo-speak for 'Aren't foreigners hilarious?' and what respect I had, years ago, for Clive James evaporated when he started to milk this peculiarly unpleasant vein. There was a moment in one programme when Clive invited us to snigger at the prizes in a Nigerian TV gameshow, books rather than family runabouts, and I've never watched him since.

I rather suspect that at 8 p.m. on Wednesday evenings I shall arrange to have myself lowered to the bottom of the well in the garden rather than watch *Oddballs*. We don't have a well at the moment, but it shouldn't take me long to dig one.

While I'm in a sour mood, let me point out that Radio 1 will shortly be celebrating the 35th anniversary of the first appearance of folk singer Joan Baez at a Newport Folk Festival.

This seems a rather curious anniversary to be celebrating on national radio, but I don't doubt that the programme will be beautifully made and, in its own way, very interesting. I don't say this solely because its producer, Kevin Howlett, operates out of the office next to the one I use at Radio 1. Well, I do actually.

Joan Baez was, is and always will be one of those performers whose suffocating righteousness fills decent folk with a nameless fury, rather like those Volvo drivers who used to drive about with 'Stop the Bloody Whaling' stickers in their windows. Now, I am as opposed to whaling as anyone and have for years operated a ban on Norwegian goods because of the Norse enthusiasm for slaughtering whales. This is, I concede, rather like my giving up smoking for Lent, a sacrifice made easier by the fact that I don't smoke, because I don't recall ever having been offered Norwegian goods. But those 'Stop the Bloody Whaling' stickers wilting in Volvo windows made me feel my life should be given over to chartering whatever kind of vessel you need for whaling – could it be a whaler, I wonder? – and getting out there and mixing it with the likeable mammals until the heaving seas run red.

Anyway – Joan Baez's appearance at the head of a demonstration singing 'We Shall Overcome' has been guaranteed to have me scratching through the phone directory anxious to join whatever it is she is protesting at.

# Old Bill

*Sounds*, 26 August 1978

---

ONLY ONCE have cell doors clanged behind your Uncle John – or, to put it another way, cell doors have clanged behind your Uncle John once only – or, to put it another w . . . (*get on with it, Ed.*) Back in 1965, as Beatlemania swept the Americas, I spent seven fretful hours in the poorly appointed drunk-tank of Dallas County Jail.

The twenty or so burly misfits with whom I shared this accommodation seemed, to a man, to be suffering from acute distress of the lower tract, and many of my new-found friends were having difficulty in keeping a recent spaghetti lunch within the confines of their bodies. The resultant odours were not agreeable. I was not myself drunk, but it was the custom at that time to place those believed guilty of the more heinous motoring offences in this hostile environment. Although my period of detention was but brief, it was long enough to make me resolve on a lifetime of model citizenship.

From time to time I have, I fear, fallen below the very high standards I set myself, but have always managed to steer clear of the law on these isolated occasions. I would like to be able to establish massive street credibility – as Steve Jones did in last week's *NME* – by listing a bewildering succession of undetected crimes, but apart from once removing a small packet of used

Ugandan stamps from Woolworths and having driven sometimes at speeds above the legal limit, I have led a blameless life. Therefore I have taken stories of police hooliganism, as outlined in the radical press, with several pinches of salt, until last Saturday, that was.

I had travelled to Liverpool with two friends who are QPR supporters (someone has to be, I suppose), anxious that the excellence of Liverpool FC and the general good nature of the Anfield crowd be demonstrated to them. The day was sunny, Young's chips beyond reproach, my new scarf was a delight to the eye, and all was set fair for an afternoon of rejoicing. Thus it was that we brave lads found ourselves struggling into a somewhat overcrowded pub in search of ale. Having eventually effected a purchase, we returned to the pavement outside to consume same.

Now, I've been in this pub many times and have often taken my beer outside when it has been particularly crowded. Never previously have the police suggested that this common practice should cease and that drinkers should remain within, although I don't doubt that there are bylaws which give them authority to do this. However, last Saturday a mounted officer decided – who knows why? – that the drinkers on the pavement should be herded back into the pub. So he started herding. At first good-humouredly, then with increasing ill-temper. Eventually he threatened to detain anyone who continued drinking in the street, and a vanload of gendarmes appeared in case he should have to suit action to his words. One of my comrades pointed out, lightly enough, that this enforced crowding in the pub was likely to be in contravention of City fire ordinances, and this seemed to enflame the horse-borne one to unreason. He backed his mettlesome nag into the crowd, physically forcing us back inside, ignoring the resultant heavy toll of glasses and bottles. Having finished my drink I slipped outside again, and when challenged by the cavalry pointed out that as I was no longer drinking I could surely remain where I was. He told me, rather forcefully, to leave the pavement entirely, and then decided to clear 30 to 40 feet of the street for his own exclusive use. To this end he started clattering up and down at some speed, and when some poor bloke ambled out of the pub and on to the freshly cleared section of highway, grabbed him by the scarf and hoisted him upwards in a distinctly unfriendly fashion. At one stage our brave cossack was even attempting to ride his horse up the steps at the back of the pub, ignoring the yells of those being hurt in the resultant crush. Thus it was that a section of pavement was cleared, much broken glass was scattered about, and several hundred people went into the match thinking dark thoughts about our brave boys in blue. As I waited for the kick-off I wondered how this same intemperate officer would have treated some poor kid dragged in for nicking a car or robbing a parking meter. Relatively small though the incident was, such things can do little to cement a loving relationship between police and public.

A Liverpool performance filled with a brooding menace that should strike terror into the hearts of those pretenders who seek to dislodge them from their pre-eminence in English football restored my good humour; and it was a contented Peel who set his sails for the M6 and the long drive home. Stopping some hours later at the Blue Boar, we pulled up alongside the Spurs team coach, and, I'm afraid, we acted like everyone else present by scampering round the coach trying to spot Villa and Ardilles. When we passed the same coach further down the M1 and waved at the man Ardilles (who, with the beard he seems to be growing, looks slightly like Pete Townshend), he responded with that well-known Latin American gesture, the thumbs up. We were all as chuffed as schoolboys. With the return of football the blue litmus paper of life is turned red again.

The Pig has just pointed out that I promised in last week's paper not to write about football. She is right too. Next week I shall almost certainly tell you, unless our Editor has other plans for me, of the

Reading Festival, where, as usual, tens of thousands of fresh-faced young people will be waiting for a succession of noisy and irresponsible bands to cease their dire racket, in order that they may thrill to my reading of the traditional 'Will Steed and Peel meet Felicity and Moira outside the *Sounds* tent with the insulin?' messages. Our pulses are already racing wildly at the thought.

# Roy Orbison

*Observer*, 30 October 1988

THERE CAN BE few sights or sounds in popular music more chilling than those of old favourites being patronised by younger musicians.

You call for an example? Well, I have always been much affected by the guitar stylings of Duane Eddy, whose 'Yep' or 'Forty Miles Of Bad Road' is guaranteed to get me into the saddle and about the affairs of the day with a song on my lips.

Yet last year Duane went into the studio and allowed himself to be photographed with George Harrison, Ry Cooder, Jeff Lynne and – as we pop writers are, I think, obliged to say – no less a personage than Paul McCartney. Worse than this, these people were permitted to play on Eddy's LP. The results were pretty beastly.

Another serious hero from roughly the same era as Duane Eddy is the peerless Roy Orbison. I first saw Roy (and I am going to have to get all folksy with you here) on stage at the Texas State Fair in Dallas in 1960. He did not quite come out, sing a song or two and advise that there would be more of the same inside upon payment of a slim dime, but the spirit of the event was pretty close to that.

Roy was, of course, splendid and every time he came to Dallas during the four years I lived there, your correspondent was first in the line for tickets.

So it was that my ample bosom was heavy with anxiety when Orbison played Ipswich two seasons ago, accompanied by a parcel of cutpurses with permed hair who introduced his set with ten minutes of wildly inappropriate ambient piffle. Roy was made manifest, burst into 'Only The Lonely', got it horribly wrong, and we settled back in our seats for some serious squirming. However, the great man plainly needed one song to clear his tubing, for from song two onwards he was wonderful – so much so that when he strummed the introduction to 'Running Scared' (the song in which, you will recall, the loser wins) I burst into a torrent of tiny, tinkling tears.

You can imagine therefore with what foreboding I approached a new LP on which Roy Orbison is teamed not only with Jeff Lynne and George Harrison but also Tom Petty and Bob Dylan, the quintet masquerading as the Traveling Wilburys (sic). Yet, by the close of side one I could almost have forgiven them all the Wilburys business. My ill-temper was diminished by an amusing sleeve note telling how the Wilburys came to travel so, and virtually obliterated altogether by Roy's solo, 'Not Alone Any More'.

Not that I would not prefer a new LP of new Orbisongs, as I am afraid we call them, recorded in the studio from which all famous friends – unless Roy knows Duane Eddy well – are barred, but the Traveling Wilburys are jolly enough for now. I even enjoyed a couple of the Dylan songs.

# Osmonds

## Hated in Hounslow – Loved in Prague

*Disc*, 1970–1

---

SO HERE I AM in the *Disc* office and there's no one else here. Mind you, it is still pretty early in the morning and I expect the next few minutes will bring a flood of keen, lynx-eyed boys and girls clutching exclusive interviews with Mud Terrier, lead guitarist with nude New York combo the Distressingly Bads.

Perhaps it is the season, aye, that must be it, but it has been a dour and bad tempered week to live through. With leaves stirring around in the gutters alongside the wrappers, cartons and cabinet ministers of a dying civilisation (I use the word advisedly); with mists and dark and that kind of dampness that causes your trousers, however clean, to feel as though the insides of them had been smeared with marine glue, then a sensitive youth like myself cannot be blamed for skulking along in the shadows and pulling frightening faces at children when their parents are not looking.

Two nights sleeping in a nylon sleeping bag on the floor of a flat have done little to improve my temper. Nylon has the less than endearing property of clinging faithfully to any small nick in your skin or any imperfectly cut toe- or finger-nail and making a night's sleep a very real torment.

Cats have mated on the balcony with a remarkable deal of

uproar and a stench that evokes the Pit itself. Friends, asked for a moment of warmth and/or a cup of tea, have replied 'Let's have lunch sometime.' In show-biz, faithful readers, 'Let's have lunch sometime' is roughly equivalent to 'Piss off, and leave me alone.'

On the credit side I did get to see the Jackson 5 and was proved quite dramatically wrong about them. It's always healthy for world-weary and cynical toads like myself to be proved hopelessly wrong. I had, and the tenor of my reviews will have shown it, supposed the Five to be a fairly ordinary supper-club entertainment, but this is in no way the case.

Playing for a gang of jaded old wasters at 'The Talk of the Town' they showed a freshness and energy that was remarkable, and Michael Jackson proved that he is a phenomenally talented singer. I didn't get to see the Osmonds but I suspect the Jacksons have them badly beat.

Enter stage right the glamorous Miss Russell (not the full-figured girl herself but the *Disc* personality) and she is talking about seeing Donovan on the television last night. She says that he was very good – I can believe it. It is a bitter comment on our lives and times that so little is heard of Donovan now. I'd love to hear him singing on *Top Gear* or the other programme again.

Several people have asked me whether (pause for a note for collectors: whenever anyone writes 'several people have asked me' it means 'no one has asked me but I can think of nothing else to write about') I saw the Queen when she came to open the BBC's exhibition of fifty years of One Thing and Another.

I did not and exhaustive enquiries have revealed that she didn't even ask after me. That hurt quite a lot. I've never seen a Royal Person and suspect that they're probably made of essentially the same ingredients as the rest of us. I'd like to have a chat with some of them and play them some good music – the LP I just bought by the Hoodoo Rhythm Devils would make a good start.

But wait, here comes our editor, austere and overweight, with a letter. It is signed 'A Mother Who Is Very Glad Of A Group Like The Osmonds' and is postmarked Hounslow, Middlesex. I suspect already that it is a hoax. 'I refuse to let you or anyone else [Note the seeds of fanaticism and repression] insult the Osmonds . . . as a mother I know my daughter will only be influenced in a good way if she follows them.

'How would you like to see your daughter burst into lament all because some ignorant person has written insulting things about the people she loves almost as much as her parents? Leave the Osmonds alone – or at least be honest about them – we all know that you are jealous, after what you've written you can't deny it!'

I bet you're thinking I've made all of this up – but I really haven't. If you criticise the Osmonds, Tom Jones, The Moody Blues, just about anyone – you're bound to get letters from irate fans claiming that you're jealous of their heroes. It's remarkable the consistency of that claim. By the weird logic that criticism equals jealousy then A Mother Who Is Very Glad Of A Group Like The Osmonds must be jealous of me. If you'd like to come and take my place in that godawful nylon sleeping bag on the floor, dear woman, then you are more than welcome.

Damn! I thought that was a relatively neat way of ending the column but the Editor, shrewd and hard-headed poltroon that he is, warns me that I have not written enough and if I think I'm going to get away with that sort of thing then I have another think coming.

So here is another letter, this one from a Vitezslav Kotik of Prague. Mr Kotik's name is covered with accents and strange devices but there is no facility for reproducing them with this typewriter. He says (and I'm not making this up either), 'All I can tell you is that I like *Disc* very much indeed. I read an each copy I get rather from the first page to the last one word by word, and what I like to read especially, that's the Mr John Peel personal column and his reviews on new singles. I'm really far from using any cheap words of praise, also surely I'm not the first person ever finding out how fine publicist (?) Mr PEEL is, but I must admit I've never met someone in the field of music press writing so intelligently and with such a large outlook on contemporary music as Mr PEEL has. That's why I wondered if it would be possible for you to send me any photo of Mr PEEL, 'cos people like him don't appear in your colour posters so often – if at all.'

Now that's much more like it – and thank you very much Mr KOTIK. I may be painfully under-appreciated in Hounslow but I have a friend in Prague. I will show your letter to the Editor and suggest that it is high time there was a colour poster of myself and the other *Disc* writers. I will even wear eye make-up and glue snails to my cheeks if it is necessary.

Next week, in response to numerous requests (which are surely not spelled like that), I will write about something. Perhaps 'The Influence of Aristophanes on the Pre-Raphaelite-Rock phenomenon' or 'Drums, Drugs, Dames, Degradation, Dropsy and Dalmatians'. Are you looking forward to it, Mother Who Is Very Glad Of A Group Like The Osmonds? And while we're at it, why 'A Group LIKE The Osmonds'? What's wrong with the Osmonds themselves?

Actually I wasn't going to mention it because I don't like being bitchy but the good lady also mentioned my 'shear ignorance'. There's one for sheep fans to ponder.

# Osmonds 2

*Sounds*, 15 December 1973

---

I HAVE HERE, clutched in my moist and oily hand, a letter from an Osmonds person. I will be brutally frank and tell you that it is by no means a cordial letter. It comes from Fan No. 2325. It would be betraying a confidence to reveal Fan No. 2325's name but I will stand up in front of you and clearly and boldly recite her letter.

It runs 'To Mister perfect Peel, (concerning your so-called review on Marie Osmond's "Paper Roses" single). Okay so you don't like the Osmonds, but do you have to refer to Marie as "Pudgy faced as her notorious brethren". The Osmonds are perfect in every possible manner. I suppose your types are hairy and ugly (like yourself). That sort make me sick, you're all hair and mouth.

'The Osmonds would never be jealous of you, so don't bother being jealous of them. And don't go citisising (sic) Donny and Marie releasing oldies, us teenagers didn't here (sic) them when they were first released.' Written in very heavily at the bottom of the letter are the words 'Hate' and 'War', and there is a poorly drawn swastika – a real swastika and not the reversed Nazi one. After some deliberation our panel of fan-letter judges have given this work 6 out of 10. This relatively low mark comes as the result of the appearance in the dying moments of the epistle of that great rarity, a valid point. Miss 2325 is quite right when she observes that the ancient songs the Osmonds choose to croon are brand new in the ears of their fans. Nevertheless, this doesn't provide the homely little fellers with an excuse for singing the luckless melodies so poorly. 2325 scored points all over the board for 'so-called review' and the suggestion that I am jealous of our prosperous wee chums.

The Swiss judge gave her an 8.5 for these and if the French had not marked her so lowly then 2325 might even have scored a 7 and qualified for the semi-finals in Rome in March. Degree of difficulty was 4.5. It is comforting for us to know, following our depressing failure in the World Cup, that our British Under-14 Irate Letter-Writers Team is producing scribes to match, in both technique and sheer guts, the enraged Elvis and Andy Williams fans in the senior squad.

I have read criticisms that the Under-14s are better at spelling but this I believe to be merely the scaremongering of embittered cranks. Some of these youngsters have it in them to achieve the stature of such legendary letter-writers as the demented vicar in Wiltshire, who suggested that I should be castrated and deported for allowing 'obscene silences' on a programme I once did with John and Yoko Personage.

I would like to take this opportunity to encourage 'ZOUNDS' readers to write insulting letters to this column. Perhaps we could award a monthly prize for the Most Illogical and Infuriated Letter of the Month. Such a contest could well brighten what promises to be a peculiarly ugly winter.

One paper I was having read to me this morning carried the suggestion that petrol-rationing could very well last for four years. If it does then the whole face of rock as we know it is going to be drastically altered. Rather opportunely I've just been discussing the impending doom on the phone with Jake Scott of Writing On The Wall.

Jake reckons that if the worst comes to the worst his band could drastically cut down on the equipment they use and travel to gigs by train – always assuming, of course, that there are trains. I've already had to disappoint thousands of adoring fans in Scarborough because it seemed certain that even if I travelled to that balmy city I would be unable to get back. The only other gig I have lined up is for Son et Lumière in Kent just before Christmas and I'll do that one if I have to walk.

In fact, I'm already faced with the prospect of having to hitch in and out of London from Suffolk to do my programmes – you remember, the ones you never miss. A round trip by car is over two hundred miles and I have to do that twice a week. With rationing I'd be able to do it once a month. If the trains stop watch out for me on the grass verge. I'm the rather distinguished looking man with the receding hair and the boxes of LPs.

I suppose most bands simply will not survive petrol rationing and the loss of gigs. Solo singers and instrumentalists will be better placed but they too will be dependent on trains. Towns without train service are going to be mighty short of live music. With vinyl and paper shortages kicking the record business several inches below the navel as well, it's going to be mighty difficult for new artists to get their careers chugging along at a workable pace. Bet you one thing though. Bet those Osmonds make it through the night.

# Osmonds 3

*Sounds*, 1 December 1973

---

THERE'S A VESTIGE of a chance, furry friends, that some of you may have cycled along the A30, the same A30 that moved Hengist and Horsa to tears so many years ago, and cursed the long hill that climbs south from the roundabout in what could easily be Egham. As you reached the top, sweating profusely in that embarrassing way you have, you may have observed to the left, through the trees, a magnificent Victorian wedding-cake of a building, the rich and quite excessive ornamentation of which makes St Pancras station look like a scout-hut.

This fine establishment is the Royal Holloway College and last week I steered my thrusting van thence at the recommended 50 miles an hour. My destination was the School of Music, an altogether humbler building on the other side of the main road. Here I was to take part in a discussion with some thirty or more souls on the general theme of 'Rockaboogie, What Now?'

I had been advised that the thirty or more souls mentioned above were more involved with the history of 'serious' music than with the actual playing of instruments for profit or gain. I had anticipated finding row upon row of dust-covered spinsters and silent, preoccupied men with copies of *Haydn, His Thoughts on Counterpoint* jutting out of the pockets of their serge blazers.

In fact, the customers were for the most part congenial and

receptive. One lady caused a brief furrowing of the brow when she observed that the only popular music with merit, in her estimation, was the music from such entertainments as *Godspell* and *Jesus Christ, Superstar*. However, we Peels are nothing if not men of valour in the face of adversity so we plunged straight away into a discussion of the Osmonds. Rather alarmingly everyone seemed to have heard of the little chaps.

We at the Peel Research Foundation have lately arrived at the conclusion that the Osmonds phenomenon, a phenomenon which consistently clogs the charts with unlistenable records by every Osmond known to a suffering humanity, is the result of a marketing exercise similar to that undertaken in recent years on behalf of vaginal deodorants.

For billions of years women have pottered through life minus the dubious benefits of vaginal deodorants and seem to have thrived on the experience. Without wishing to plunge too deeply into the biology of the thing, and thus draw artillery from Deeply Offended of Grimsby, menfolk seem to have been quite contented with the existing arrangement too.

However, a shrewd businessperson murmurs *sotto voce* 'Are you causing offence?' and hordes of women dart out to flavour themselves with attar of roses and/or roses of Picardy. I like to feel (whenever I can) that the human race would have survived in its thrill-packed dash to oblivion without vaginal deodorants – and without the Osmonds too.

We discussed this point at the School of Music for some time. We also discussed earlier phenomena like Frank Sinatra, Elwood Pretzel and yer Beatles. The consensus was that these latter were a good thing. From here the debate moved to weightier matters and I found myself trying vainly to explain why I like the Faces, the Floyd and The Who, but not ELP, Yes and Focus. I was grateful that none of the students brought up any of the technicalities of music; I wouldn't recognise a flattened fifth if someone slipped it into my breakfast foodstuffs.

It came time to illustrate some of my non-points with a recorded work or two and rather opportunely we started with a spot of Tangerine Dream from their admirable double-LP *Zeit*. This (and other Tangerine Dream records) is on the Ohr label and can be had off the better import shops. If there be any among you who have yet to hear Tangerine Dream, then hurry, hurry now, while the offer lasts. They are spectacularly good.

The students of the Royal Holloway College seemed impressed and several mentioned that *Zeit* bore more than a passing resemblance to the efforts of several European 'serious' composers. (Which is not to assume that T. Dream are in any way frivolous.) They mentioned in particular Ligeti (a fine sweeper, who only last Saturday saved a point for his side with a daring goal-line clearance) and older readers may recall that some of his stuff was used in the soundtrack of *2001*.

Maddened with success I moved over to some of the other records in the grab-bag of goodies I had carried with me. I think I made a mistake in so doing, should have quit while ahead, no judgement etcetera. Before we started it was agreed that anyone already bored should leave and about four-fifths of the audience scampered into the night air.

For the gallant remnant I played the Cougars' 'Saturday Night At The Duck-pond', a 1963 Shadows-type single based on *Swan Lake*, and argued without convincing anyone that this was a worthier treatment of the classics than that meted out by ELP. The gathering were politely interested in the Floyd's 'Interstellar Overdrive' and, most surprisingly, enjoyed Gene Vincent's magic 'Jump Back, Honey, Jump Back'. Interest flagged considerably when we turned to 'Bal Pour Un Rat Vivant' by the superior French group Komintern and 'Time Captives' from the defunct Kingdom Come. During these works the murmur of conversation became a hub-bub (that can't be how to spell it) and probably centred on Rossini's ability as a short-order cook or something equally historical.

Interest revived briefly with Nigeria's Oladunni Decency and Her Liberty Orchestra but I was beginning to feel freakish and ill-at-ease. The room was gradually filling with boys and girls who were observing proceedings with the detached interest of children confronted with the World's Fattest Man or the Amazing Bearded Lady. After trying to recover my initial successes, with records by Linda Jones and The Original Blind Boys of Alabama, and meeting not acclaim but a frosty silence, I reached for my hat, made my excuses and fled into th'encircling gloom in search of petrol.

Next time I'll stick to the Osmonds and Tangerine Dream. One knows where one is with the Osmonds, doesn't one?

# The Party's Over

*Sounds*, 10 January 1976

HANDS UP anyone who can tell me which record is currently holding down the coveted 191st spot in the American album chart? No guessing now! I'll tell the rest of you in a moment or so.

I know that you out there believe that I spend too much of my time bleating feebly about the shocking suffering we parasites on the rotting time, a time made up of equal portions of strong drink, exhilarating substances smuggled in from the world's most remote provinces, young women (or men) skilled in the subnavel arts, and perfervid evenings at the opulent homes of the stars featuring all three.

Ah me! If only it were so.

Instead we have to swallow our pride and go instead to record company Christmas boozeramas.

These take several horrid forms, but it is the principle of the things that you arrive – even if you arrive at dawn – several hours after everyone save the employees of the company has left.

In consequence, your arrival interferes with a number of long delayed and elephantine courtships between promotion youths and overstimulated secretaries liberated by gin.

Conversation is entered into only with the object of getting you off the premises as quickly as possible, and drinks are mixed with the same motive and in such immoderate measures that it is clear that the amorous youths are prepared to risk hospitalising you to clear the way for their ugly wooing.

There are occasionally exceptions to this melancholy principle – one record company came up with an interesting variant this year when stunned revellers, such as myself, were treated to the nightmarish uproar of a scratch combo of which Signor Paul Gambaccini on piano was quite easily the most accomplished member and which featured radio personality Tony 'Your Royal Ruler' Prince chanting rock and roll favourites with admirable confidence and in a colourful range of inaccurate keys.

From this curious entertainment my companions and I hurried across the hurly burly of London's brilliantly illuminated West End to A&M Records' Party at that end of the King's Road at which it is possible to buy something other than shoes and boots.

A&M's was the best of the Yuletide festivities I witnessed – even Bomber Harris was, I am assured, numbered among those present – and I was just demonstrating to a room crammed with admiring celebrities exactly how it was that I foxed the German High Command back in '43, when it was pointed out to me by a wellwisher that I was due to drone on the nation's airways in a matter of minutes.

Unfortunately I had dismissed Spring, my chauffeur, and was only rescued in the nick of time by the offer of a lift to Broadcasting House from young Andy Fairweather Low.

On our drive I discovered that Andy is, like myself, a great admirer of the earlier work of Jimmy Reed; that he did, in truth, have much of Jim's Vee-jay material on 8-track to brighten the tedious drive down the M4 to the land of his fathers.

I was so moved by this and by Andy's generosity that I gave him one of my most treasured 8-tracks, that by the Kate Brothers. I hope it pleases him as it pleased me.

I know perilously little about the Kate Brothers, I am afraid. Their album, their debut album, arrived at Peel Acres in the same airmail packages as the record which is some 180 places ahead of it in the American chart, *The Hissing of Summer Lawns*.

From the sleeve photograph Mr and Mrs Kate's little boys look more like long-distance truck drivers and the kind of Californian hillbillies who usually bubble to the surface on *Asylum*.

Their music I recommend quite earnestly and I am pleased to see that some of New York's discos have picked up on 'Union Man', from the album, and are playing it nightly to their widely rejoicing customers. 'Union Man' has a pair of the niftiest little instrumental breaks around town.

# Peel and the Mighty Gorgon

## A Gripping Tale for Boys (and Girls)

*Sounds*, 8 November 1975

---

WHAT A DULL column that was last week, eh? I'm most dreadfully sorry about it. Here's a bit of action to make up for it.

Ned crept across the rocks towards the smugglers' lair. Close behind him moved the silent figure of his new friend Earl, the American. The intrepid pair could see quite clearly now the pinprick of light inside the cave.

'I bet they have a schooner anchored off the shore somewhere,' hissed Ned. 'Sure thing,' affirmed Earl, who was still ruefully rubbing his damaged wrist. 'I betcha Vombini and his henchman will make a break for it before dawn.' At that moment Ned's foot slipped on the rock, dislodging a stone which rolled noisily towards the mouth of the cavern. 'Gosh! That's torn it,' thought Ned.

Pretty gripping, isn't it? I do like a good yarn, don't you? Now here's a true story. Are you sitting comfortably? Then let's change places. (The old ones get 'em every time.)

'John Peel's here,' announced the man behind the counter. From the back of the shop and out of sight came guffaws of laughter. Your hero shuffled nervously from foot to foot. He had grown accustomed to people yawning rather pointedly when he was announced, but gales of unrestrained merriment added up to a new experience. I bet you're just dying to know where I was. Well, if you'll sit still and stop fidgeting, I'll tell you.

Once a week I take the tube to Victoria Station, in the heart of London's bustling, colourful, West End, then change on to the Southern Region and journey to Peckham Rye. Once there I amble through the streets until I come to the doors of Intone Records. This is where I spend far too much money on reggae records. The noise within tends towards the deafening, with extracts from current releases being played at considerable volume and without pause. This, although it means that all communication must be by hand-signals and shouting, helps the customer to decide which records to buy.

The end result is that I stagger back into the street with my nerve ends jangling, my throat raw and my weekly allowance sorely depleted. I suspect that those who work in the shop regard me as being seriously disturbed – hence the uncontrollable mirth – but they are always helpful and have yet to recommend a duff record.

This week, for example, I bought an LP by the impressive Burning Spear. This is on the Fox label and is called *Marcus Garvey* but, unfortunately, the copy I have is pressed slightly off-centre so I cannot play it for you on the radio. They also brought to my attention the classic *Mighty Gorgon* (on Klik KL 607) by Cornell Campbell, which is currently one of my most valued singles. Intone Records assure me that they can take mail order business, so if you're interested their address is 48 Peckham Rye, Peckham, London, SE15.

While I'm involved in giving blatant plugs to record shops – and, unfortunately, none of them bribe me to do it, nor do they give me a discount on the records I buy (and why should they?) – why not try Black Wax at 12 Mitcham Lane, Streatham, London, SW16 6NN? I'm often to be found parking illegally outside their premises while I devour my substance on exotic

American soul singles. From them came such gems as the Bronner Brothers' 'Hold On To God's Unchanging Hand' (Jewel), 'Closer To The Aisle' by Esquires Ltd (Smog City), and Papa and the Utopians' extraordinary 'Cry For Joy' (Poizon), all of which you have doubtless heard and enjoyed on my radio programmes.

If you continue to listen you will, in due time, also hear 'Mr Independent' by the Soul Twins (Back Beat), 'My Love For God' by Tommy Ellison and the Five Singing Stars, and 'It's Been A Long Time' by Ray Williams (Vasko). These also come from Black Wax. I wonder if it is records such as this that have caused a truculent listener in Leicester to write and say: 'I am just writing to say how totally disgusted I am with the music you now play. Here's hoping that you come to your senses in the near future.' If it is then I shall send Ned and Earl, along with several other Peel Foundation heavies in military-style uniforms, to rivet the offensive and offending writer's head to a double-decker bus (42 seats, 8 standing. No spitting).

Next week I may tell you about The Members, J. M. Lulu, Mountain Sisters, Ladysmith Black Mambazo, Reggie's Soweto Magic Band and the Love Birds and The Creations. Also the Mahotella Queens, whose singles, on the Gumba Gumba label, are rousing indeed. Now I bet you're thinking, 'What an old silly that Peel is, to be sure, making up those funny names.' Well, boys and girls, I have not made them up. A man called Jumbo, an operative with Virgin Records, has sent them to me and if I wasn't so butch and manly I'd rush round and plant a great big wet kiss on his ivy-mantled cheeks.

Also next week – Genesis, The Truth, By One Who Knows. Maybe.

# Peel at the 'Quiet' Albert Hall

*Disc*, 31 July 1971

THE PIG had been abed for a day or so, felled by some mysterious ailment that had left her weak and depressed. After *Top Gear* I decided to take her to a concert at the Royal Albert Hall, feeling that some good music would make her feel a lot better. Having phoned to reserve tickets we set off to pick them up. The man on the door we asked for information was breath-takingly rude and we stood, chastened and dispirited, in the queue, watching another doorman bullying the people who were filing into the arena part of the hall. 'He's shouting at them as though they were schoolchildren,' said the Pig.

The doors opened at 7 p.m. and we found our way to our seats. They were high up behind the stage and when the organ was played we both vibrated. It felt pretty good too – don't know whether they were worth eighty pence each though. Already the people standing about the footlights were making a lot of noise and we sat forward in our seats expecting to see some of the unruly behaviour that seems to get groups banned from the Albert Hall with such monotonous regularity.

To our horror the fans were shouting abuse at members of the staff who were, after all, only doing what they're paid to do. In fact, they weren't just shouting individually but there seemed to be a number of ringleaders who stirred them into chanting abusive slogans. We noticed several of the more active individuals hurling missiles on to the stage and we both strained our eyes peering into the back of the hall for a sign that blue-uniformed retaliation was about to descend heavily on the miscreants. Astonishingly it never came.

Possibly the rowdies were encouraged by this official turning of the blind eye because disruptive elements began to single out eminent members of the audience for special attention. 'Hello, Richard Baker,' they chanted. 'Hello, Robin Day.' Then they sang 'Why was he born so beautiful, why was he born at all?' and came the answer, swift as thought, 'He's no bloody use to anyone, he's no bloody use at all.' 'That's it,' said the Pig and I in rather harmonious unison, noticing that the whole deplorable episode was happening in front of the BBC2 TV cameras, 'this time they've gone too far.' But there was still no retaliation.

A man, who could have been the manager of the Albert Hall, made a few opening remarks. To our surprise he seemed to thrive on the repeated interruption and disorders from around him and he finally left the stage to cheers. As he went the musicians came on and while they tuned their instruments the noisier spectators hummed along with them and everyone in the arena seemed to be hushing everyone else. Obviously, thought I, this sort of undignified conduct will, sadly, lead to the banning of yet another group from the Albert Hall so, if you read the Royal Liverpool Philharmonic Orchestra have joined The Nice, Deep Purple, Frank Zappa, James Brown and others on the black list, I want you to know their fans are to blame for it.

Well, let's face it – you and I know that the Liverpool orchestra isn't going to be banned at all and we also know that the uproar and lack of restraint at the last night of the Proms will at least equal and probably exceed anything that's ever happened at a 'Pop' concert at the Albert Hall. Obviously then it's not what's done that matters, but who sees it.

The arena full of studenty types was flanked by rows and rows of neatly, conservatively dressed middle-aged folk who smiled indulgently, even fondly, at the display of youthful exuberance before them. These were their children enjoying their music – would they have smiled so securely, even smugly, if their children had been reacting in a comparable manner to Mott the Hoople instead of the Royal Liverpool Philharmonic Orchestra? I suspect not.

Pig and I were in a poor mood before the music started, enraged by what seems to be a clear double standard being applied by the Albert Hall's management. Audiences at what we'll call, for argument's sake, 'Pop' concerts are no more aggressive than those at the Proms. Both groups of people (and there are probably many who would be part of both) are there to enjoy themselves and react to the music. The only unpleasantness I've ever witnessed in the hall was in response to a hysterical over-reaction by a member of the staff at a 'Pop' concert, to the kind of audience reaction at which everyone was smiling last Saturday night. What's sauce for the goose is clearly NOT sauce for the gander. There's one law for 'classical' fans – another for 'pop' fans.

Our rage was abated somewhat by the excellence of the music. Sibelius' Sixth Symphony is a soothing work full of rhythm and clear, open vision. The programme notes called it 'unemotional' but that's not the way it reached us. This was followed by Grieg's Piano Concerto in A Minor. This is probably regarded as a much over-performed work by the cognoscenti but we were delighted to hear it again. Pianist Radu Lupu was fine visually as well as musically, boasting a great mass of hair and a beard, and at times he bore a strong resemblance to the majestic Captain Beefheart. To watch him, at the end of a piano passage, hurl his hair back and turn to watch, glowering, as the orchestra played their part, was worth the price of admission in itself.

The audience continued to cry out from time to time, shouting such inexplicable things as 'Squeeze' and 'Heave.' The only glimpse of humour, as opposed to conditioned response, came when several people shouted 'Will the real Bill Shankly please stand up?'

Half time activities were reminiscent of the struggle to get into the approved toilet facilities at Anfield and we were glad when we were able to settle down to hear Naylor's specially commissioned 'Scenes and Prophecies' which started the second half. At times it reminded me irresistibly of 'Atom Heart Mother' and contained the following revealing quote from the prophet Joel. 'Gather yourselves together' – doubtless at your little cottage in the desert. The composer emerged to acknowledge the applause at the end and looked like a nice man, which made the evening feel better.

The Scouse Orchestra seemed happiest with the final work, Elgar's 'Enigma' Variations, played it beautifully and obviously enjoyed it greatly. When it had finished the audience reacted very strongly indeed and were still cheering and roaring for more when we left.

'I expect they're all drugged,' said the Pig.

# Peel's Theory

*The Listener*, 8 March 1973

I'M SURE that any theory, if it is to be presented to the public, should be carefully considered, debated and examined before it is so presented. An hour ago I was going to write a piece about German rock, a piece crammed with obscure names and an enthusiasm for long and meandering album tracks. However, I went instead to a pub and read a few of the letters that come to me at the BBC. Now, in addition to presenting a few radio programmes for Radio 1, I review the new pop singles for one of the national music papers. Naturally enough, I do not choose the records I review and the editor insists that I cover releases by such current teen favourites as Gary Glitter, various members of the distressing family Osmond and David Cassidy. I do, I will confess it, find it hard to be fair about records by these and other heroes, yet, as a former Lonnie Donegan fan, I know how it hurts to read cruel and violent reviews of the works of your favourites. So I try to be fair and objective and to review these singles in their own context.

No matter how hard I try to be fair, though, I find myself on the business end of a torrent of extremely abusive letters and it was during the reading of some of these, less than an hour ago in the pub, that my theory began to take some sort of shape. You may recall that when the Rolling Stones first publicly practised their particular art they were subjected to a great deal of abuse of the 'why don't you get a hair-cut and have a wash and get a proper job?' variety. These and similar questions are being asked by my irate pre-teen correspondents. They attribute my lack of appreciation for the work of Mr Glitter to a wholly imaginary drug habit. My apathy towards the Osmonds both individually and en masse they feel certain is due to my reluctance to bathe. My failure to appreciate David Cassidy is due to the length of my hair, the regrettable fact that it's receding apace and the curious sex life that they generously attribute to me. The point is that today's ten-year-olds are slinging the same mud at us that our parents threw about wildly ten years ago. Therein lies the seed of what may be known, if you insist, as Peel's Theory: that is, that the right-wing backlash comes not from the Longfords and Whitehouses but from below – from the Osmonds' fans.

Further support for this sophisticated theory comes from the correspondence columns of several papers that would a year or so ago have been described as 'Underground'. In one letter a mother, having praised the recent triple LP from the Grateful Dead, bemoans the fact that her five-year-old child is extensively 'into' Donny Osmond. 'Where,' she cries, 'have I gone wrong?' In another, a young reader leaps to the defence of David Cassidy and slams 'filthy, long-haired groups' for their drug habits, their sexual mores and their lack of familiarity with soap.

The music of these crypto-stars, particularly Donny Osmond, is distressingly reactionary as well. Gary Glitter, admittedly, bases his work on rock 'n' roll, but this is done in such a way that the essence and power of the original is ruthlessly expunged and we're offered a substitute that is all packaging and no substance. When I tell you that the latest offering from Donny Osmond is a bland and tasteless re-make of Johnny Mathis's hit 'The Twelfth Of Never', you'll see what we in the middle are up against. If you have a son or daughter, brother or sister who is 'into' the Osmonds, watch out: they may turn you over to the police for not having a television licence.

# Pen Pal

*Sounds*, 24 April 1976

---

MY FANS – and you know I love every last one of you, you little sweeties – have been at it again.

Every once in a while as Fifi and Yvette, my secretaries (both models of rectitude and virtue), sort through the tear-soaked requests for a lock of my hair (and you thought my incipient baldness was due to age, didn't you, you tease?) or for the address of my tailor, they stumble across a letter or card written by some unsung philosopher or sage, some latter-day Capt. W. E. Johns.

This very week Yvette brought me a letter from David, who forgot to state his address, who has some jolly telling points to make about one of our favourite combos, Emerson, Lake & Parker.

What David has to say is, I believe, rather important, so I intend to bring you the uncut version of his letter, which is headed, rather obscurely, 'Death to Reggae Musicians'. Wasn't it the Minister For Agriculture & Fisheries in the post-war Attlee government who said – and I think quite rightly – that 'Death comes to us all, you know'? I've always thought that was rather beautiful. I'm sure you agree.

'Dear Shit', David starts his letter (that's how Yvette knew at once that it was for me), 'Emerson, Lake & Parker, eh?' Then my correspondent goes on to say, 'Ya shit-faced (and here follows a word used fairly frequently to indicate one of the more sought-after portions of the female anatomy). Your persistent slagging-off of the greatest group in the world pisses me off. Why don't you take a look at your own taste in music before you slag off others? The very mention of your name causes me to vomit.'

I think we'll pause here for a moment. I'm most interested to hear of the effect the words 'John Peel' have on David. One or two of my closer friends have told me, in the strictest confidence, that they find David's claim rather hard to believe. I reserve judgement.

David continues, 'You're a fuckin' perverted bastard' – can't argue with that, old son. Then – and I think this next bit is rather poetic – 'You remind me of a dog's crap that's been lying in the sun and gone all white. As for your DJing qualities, well, they're non-existent. Freeman compared to you is like ELP compared to the Faces and Little Feat.'

Now I have a lot of respect for Alan, I think he's a marvellous old man and a shining example to us all, and I think David is being rather unkind to him making this comparison.

Back to Dave – 'Also your nauseating arrogance makes me want to piss on you, ya clit. You're also a crawling insect, I mean look at the way you crawled up Phil Collins' arse. For years you slagged off Genesis but when you managed to get Phil on your crappy prog it was Yes Phil Sir, No Phil Sir, but of course the album's Superb Philip Sir CRAWLER.'

Well, Dave, it is true that I have never been a great admirer of the work of Genesis. However, I thought *Trick of the Tail* better than any of the previous offerings from the band, and thus we asked Phil Collins to saunter along and tell us about it on the 'crappy prog'.

Not caring for someone's music doesn't necessarily mean, Dave, that you don't like the people concerned themselves. I have, for example, a lot of time for Keith Emerson, who is always most amiable on the rare occasions when we meet, and for Rick

Wakeman, who is as entertaining a layabout as one could hope to meet. On the other hand, one or two of the people whose music I do like are complete twerps.

As a result of the man Collins' coming into the studio we booked his alternate band, Brand X, who were, I thought, excellent. But let's return to the ruminations of our letter-writing pal.

'One of my acquaintances absolutely idolises you and listens to your waste of 1 hour every night, and he's a bastard too.' We bastards must stick together, eh?

Next Dave asks a question, 'So why don't you take a trip and go and meet Duane, Jimi, Janis, Brian, Paul, Berrie you necrophiliating spunk bag? Away and wank. Yours, with deep hatred and the utmost animosity, David.'

My goodness! How the human passions can be stirred by differences in taste. Thanks for your letter, David. If you'd care to write again and give me your address I'll try and write you a slightly less hysterical letter than your own, explaining my point of view on ELP and related matters.

Incidentally, I do know that the band is called Emerson, Lake & Palmer rather than Emerson, Lake & Parker. My use of the latter and inaccurate name is just my tiny little joke. What a silly I am, eh?

As for the rest of you, what do you think you're doing standing around smirking? You should be out and about trying to find me a 12-seater mini-bus in good condition and at a fair price.

We need one rather badly for the small community in which I live and the only such vehicles I've seen so far have either been too expensive or too dilapidated. I know there are dealers who specialise in such contraptions, but they all seem to employ salesmen who wear large and brightly coloured ties. I never trust such people and I don't imagine you do either.

# Phoenix Festival

*Guardian*, 23 July 1993

---

SINGING AND drumming at the same time has always seemed to me a pretty neat trick, one that Sam Marsh of the Bury St Edmunds trio Jacob's Mouse has off to a T. Sam's trio was the first band to strut its stuff on the main stage at last weekend's Phoenix Festival, held in the featureless, windswept grounds of the Avon Park Raceway near Stratford-upon-Avon, and the Mouse's no-nonsense, post-grunge rock was just the ticket.

I had been hired to play records betwixt performances at the Raceway and to introduce the bands, so I next summoned up the God Machine and then Fatima Mansions, both of whom offered workmanlike if unremarkable sets. Sadly, the amusing remarks I had polished to a diamond brilliance to introduce Hole, fourth on the race card, went unheard, as I had forgotten to open my microphone. With an agility born of years of experience, I blamed this on the sound crew, thus guaranteeing myself a weekend of apologetic tea and stale biscuits.

There has been much recent interest in Hole, principally due to singer Courtney Love's keenness for winding-up the press. Courtney is also, it must be said, married to that Kurt out of Nirvana. Happily, this latter counted for little, despite some bonehead in the audience punctuating every pause with 'Where's Kurt?' Hole turned in a fervent forty minutes which our party of ten voted the best of the weekend. This is a serious band and no one-woman show.

Reason insists that someone somewhere, probably at a dull university in Minnesota, is working on a paper which will look into the reasons why predominantly white audiences prefer to be called 'motherfuckers' by white rather than by black artists. He or she would have identified much source material at the Phoenix Festival. I had planned to keep a record of how long it was before each rap-derived act so described us, but the Disposable Heroes of Hiphoprisy, following Hole, set an unapproachable standard by calling us motherfuckers on their radio mikes before they even reached the stage. Actually the Heroes were rather telling in a grumpy sort of way, as was Julian Cope. Julian performs what I can only describe as care-in-the-community pop, almost but not quite obscuring the excellence of his voice beneath layers of good-natured piffle and a range of bizarre costumes. The last of these revealed that Julian has a surprisingly pert bottom for a man of his years.

Sonic Youth brought Friday night to a close with a set which brought out the best in them – hummable tunes displayed against a backdrop of clashing guitars – and the worst – self-conscious needle noodle noo stuff which serves only as an irritant. Perhaps that is the point.

Meanwhile I was tussling with a turntable which worked only intermittently on Friday, thereby inhibiting my wizardry on the wheels of steel, and gave itself over on Saturday to a death scene of a duration rarely experienced outside the opera house. The rage this engendered gave me insight into the feelings of That Petrol Emotion who could only play for fifteen minutes following massive equipment failures.

In addition, the boys and girls of Environmental Health had complained about the noise and the level on the records I was able to play was kept so low that I could listen in on the conversations of members of the audience twenty feet below me.

In the 1970s, bands such as Yes and Emerson, Lake and Palmer were much admired for the mountains of equipment through which they played their dismal music and this tradition was revived by Faith No More on Saturday night and Living Colour and the Black Crowes on Sunday night. As I tussled with my one turntable, a large and impressively self-important roadcrew tested everything with lunatic thoroughness.

After half an hour of these equipment wars it was perhaps inevitable that the cliché-ridden tripe that emerged from Faith No More's zillion dollars of hi-tech gear was nowhere near as enjoyable as the music that, say, Die Cheerleader, the day's minor league openers, produced on one hundredth part of the resources. Mind you, the paying customers went mad for it.

Sunday started well with an impertinent 30 minutes from CNN, a trio of revivified Goths with a first-class début single in 'Young, Stupid And White', but nosedived into a sequence of underachievers enlivened only by Mercury Rev and Pop Will Eat Itself. Following their set, I was told that headliners the Black Crowes would not require my services as provider of melodies for the mudspattered. The Crowes would very likely have filled me with glee in 1973 but in 1993 they sent me scurrying to the Wedding Present and the Zine tent. The Weddoes are, for no reason I can discern, the target of the music weekly bully boys, but their modest run-through of songs set against spiteful guitaring that, were it coming from a modish US outfit, would be the talk of the town, was just what I and other members of the party wanted after a curate's egg of a weekend. As we queued to leave, someone jumped out of the car in front and handed me a note which read 'Dear John, you were a shining star in a dull festival. We love you. XXX.' Even making allowances for irony, that was just what I wanted to hear. Jo, Alex, Ant and Johnny, I hope you are going to Reading. We are.

# Pink Pop

*Sounds*, 27 May 1978

SOMEONE CLAIMING to be called Anne has written to me from Preston, only a few miles from where my saintly old white-haired mother was born, offering to fellate me until I faint. What, I find myself compelled to enquire, are today's young persons coming to? Or, for that matter, in?

When I was but sixteen years old and the sprightliest drummer-boy in the King's Dragoons, it was tantamount to a proposal of marriage if a girl were shyly to offer one a hand-picked nosegay of lavender or cherry-blossom – and a stolen kiss behind the tombstones on the way home from evensong enough to bring tearful mothers and grim, choleric fathers together to discuss whether the errant male should be transported to Van Diemen's Land, and to certain death in what Capt. W. E. Johns has characterised as 'that white Hell', or, more harshly still, to be sent for a holiday in Australia.

In 1978 a person of mature years – myself, for example – stands entirely without the protection of the law from such monstrous proposals as that from this depraved and licentious bawd. And, Anne, you poor, poor, poor broken thing, you forgot to send me your address.

I find that I must again offer apologies to my hundreds of thousands of slavish followers – this time for having forsaken them for an entire month. I must also apologise for the extraordinary photograph carried in last week's *Sounds*, in which some cruel lensman's device made me appear to be the worse for drink. (See colour section, page 2.)

Ackcherly I had rather a good European Cup Final. At three in the afternoon I was already in Trafalgar Square, a six pack of Portuguese lager fastened to my belt, with a colourful rosette pinned to my chest. Two hours later, with an entirely different six pack of lager secreted about my person, with an ill-fitting, very silly and disgracefully expensive hat on my head, and with a red and white chequered banner draped about my shoulders, I was squatting in the historic dust of Wembley Way observing, through eyes that seemed to focus but poorly, the ebb and flow of humanity.

The match itself I spent behind a lad named John – there are a great many of us about, so watch it! – who remembered that when writing in *Sounds* of Liverpool's first European Cup win, I had described Tommy Smith as a 'brick gazelle'. As His Divinity Kenny Dalglish scored at Wembley John turned and flung himself at me, with the result that we crashed to the terracing, bringing several others down with us. I carry the scars with pride.

Later, much later, I was to attempt to address a handful of Bruges supporters in one of their native languages (French), only to be forced to beat a retreat when I discovered that I couldn't proceed beyond 'Mesdames et Messieurs'. Later still I was to be found at the Speakeasy introducing the Brethren of the Snivel before they played some of their disingenuous (it is alright – I don't know what it means either) songs for a gathering of Maltese waiters and former roadies for Uriah Heep. It was then that I was photographed.

My only wish is that the lynx-eyed photog had been around earlier as I weaved down Wembley Way, banging on the windows of passing coaches with my trusty banner, beneath the benign eyes of London's finest. Now, that would be a picture for William to carry with him when he turns out at Anfield for the first time.

Speaking of Bruges, I drove within earshot of that noble city on my way to officiate at Holland's Pink Pop Festival. I never understood exactly what an English-speaking disc-jockey, whose only words of Dutch are 'Sodemeiter op' (which is very rude) and 'Krutkrekel' (which is very rude), was doing compering an event in Holland. I still don't. I can only imagine that I was the butt of a complex (and expensive) Dutch joke based around the fact that Peel (or, more accurately, 'piel') is the Dutch for 'prick'.

However, I thoroughly enjoyed the festival, at which Graham Parker and the Rumour ran out clear winners even if that Geoff Barton (who, dear Editor, I only saw once during the day and that was when he was headed for the beer tent) (*that's your fault – you should have stepped out of the beer tent more often – GB*) tells you differently elsewhere in this issue.

Several experiences at Pink Pop confirmed me in my belief that I am much better off living here in England than in my former home in California. For one thing I cannot cope with the fearful heartiness and blinding insincerity of American rockfolk.

On my arrival, with friends, at our hotel in Maastricht, I was cornered by the road manager of Journey, who somehow had got the impression that I was a sound engineer, and went into one of those 'Howdy, I'm Brad. Downright pleasured to meet ya' routines that you think exist only in the more syrupy Disney features, then insisted on introducing us all to various members of the band who clearly – and rightly – were not at all interested in meeting us.

As we finally parted after a couple of minutes of half-baked conversation and a further round of firm, manly handshakes, I observed to my friend Bert that our new acquaintance, despite his protestations that meeting us was a real pleasure – nay, an honour, would forget that we even existed as soon as he turned his back on us. Sure enough, on stage the next morning I was greeted with a cheery and apparently sincere 'Howdy, I'm Brad. Downright pleasured to meet ya, yes sirree. Folks back thar sez y'all are the compère of this here shindig.' (I may have the language a trifle wrong here, but the flavour of the thing is right enough.)

At least, as I remarked to Bert, we Brits know that when someone from our own scepter'd isle tells us that he or she is pleased to meet us, well, he or she is – or in politics. Americans, I fear, are too positive for their own good. Witness Jimbo Carter, America's first Disneyland President.

I wonder if Anne of Preston is an American. I don't faint easily, young woman.

# Please Listen to My Programme

## We've Got Jimi Hendrix, Y'know

*Sounds*, 17 April 1976

---

IN A MOMENT or so we'll have a passage from my forthcoming novel, *Buckskin and Buggery*, a rollicking yarn of redskins in the rigging.

This is the story of a little-known tribe of painted heathens of a nautical persuasion who strove in vain to drive the King's Navy out of the waters of Old Chappaquidick Bay. An exciting promotional campaign featuring bars of chocolate has been planned.

The other good news concerns Jimi Hendrix and the several sessions he recorded for the BBC. For years I had believed that these sessions had been erased and the tapes re-used for *Any Answers* and *Gardener's Question Time*.

As I'm sure you realise, tapes of these and similar programmes are kept in lead-lined containers placed at the bottom of a disused mine in Berkshire. Only authorised personnel may enter the compound surrounding the mine entrance, and unauthorised personnel are shot first and asked questions later.

On the other hand, tapes of Hendrix, Cream, the Beatles, the Stones, The Who, etcetera are either erased or chopped into tiny fragments by the laughing boys and girls of the special Culture Squad. They wear carefree black T-shirts on which 'Culture' has been misspelled 'Kultur'.

Anyway, recently some of the tapes have come to light in the storm shelter of a house near Lubbock, Texas, and we will be broadcasting them on this very Thursday, April something like the 15th.

Among the titles from which we will build this programme are the following: 'Driving South' (two versions); 'Little Miss Lover'; 'Experiencing The Blues'; 'Burning Of The Midnight Lamp'; 'Hound Dog'; 'Hey, Joe'; 'Love Or Confusion'; 'Foxy Lady'; 'Stone Free'; 'Spanish Castle Magic'; 'Day Tripper'; 'Gettin' My Heart Back Together Again'; 'Wait Until Tomorrow'; 'Can You Please Crawl Out Your Window?'; and 'Hoochie Coochie Man'. There's also a fairly bizarre Radio 1 jingle by the man.

Hendrix buffs will observe at once that there are several numbers here which have never appeared on any Hendrix record, authorised or unauthorised, and there are several others which have been heard only on rather seedy bootlegs. A treat, kats and kittens, is in store for you. If you don't listen heads will be worn bowed for years to come.

It would be nice if this material appeared on record sometime, for, as I have observed countless times on my debilitating radio programmes, even Hendrix ordinaire is well better than most folks' rarest vintages.

One wonders, doesn't one, whether perhaps such historic *Top Gear* sessions as the two Jeff Beck Group outings (with Uncle Rod in pre-Hollywood effulgence), the Yardbirds one with Jimmy Page, those by The Who and Cream, and the classic Bonzos session that gave us the gem-like 'Craig Torso Show', are gathering dust in some Corporation garret somewhere.

One wonders, and the heart beats a little faster. The Pinks did one for us too, you know. Remember 'The Massed Gadgets Of Hercules'? Ah, that Mrs Pockett from 11a. What a gadget she was, to be sure!

But I know you're waiting for a glimpse of my new book. I'll just slip out to the main hall, call Meadowes in the garage and ask him to look in the Wright-Durex V12 'Tax-Evader' for the manuscript. Talk among yourselves for half-a-mo.

Jim Cottage was helping old Nathan to splice the top fo'gallants. 'Nathan,' he cried, above the howl of the stiff sou'wester that was blowing in from the Straits of Hispaniola. 'Nathan,' he cried again, but the older man continued to finger his follicles without giving vent to a reply.

Jim shrugged and turned his thoughts to Gwen, the Captain's daughter, whom he had last seen laughing and drinking with that cur Montmorency in a low tavern in Port St Louis. How he wished that he had stepped out of the shadows as they whispered and moaned in the bus-shelter, and had hurled his defiance in the face of the caddish deserter.

What would Sir Manifold have said? And who the hell is he? And what is he doing here? Ned, sorry, Jim bit his lip as the wind buffeted the creaking timbers of the 'Golden Wonder' and thought of England, home and Mother.

I think that's enough to be going on with. You can see already, I feel certain, that this is a deeply moving work which takes the lid off the seamy world of mini-cab drivers and short-order cooks. If I can't think of anything else to write about next week there may well be more of it.

# Prince

*Observer*, 31 July 1988

---

'LIFE', suggests Cat, dancer and apparent erotomane with Prince's non-stop revue, writing in the £6 concert programme, 'can only be understood backwards but it must be lived 4 wards.' This tells you that you are getting more solid thought for your dollar here than with the sorry bulk of pop performers, amongst whom brain death often appears to be endemic. You are also, with these funsters, getting more sex – albeit pantomime sex – than at the average concert.

A fair number of the disciples at Wembley on Thursday appeared to have readied themselves more for the boudoir than for what are usually referred to as 'the concrete canyons' of the Arena, and it was refreshing to be able to peer enviously at them to elegant harp music rather than the tinny and overloud near-disco that prefaces most events here. For once, too, the stage itself was interesting to look at even before the artistes trod its boards, being large, irregular in shape and covered with such lumps, bumps and engines as suggested that this would be a performance you could not afford to take your eye off for a moment. And so it proved to be. Once the houselights had dimmed and a crescendo of anticipation had been allowed to develop, a full-sized American car was made manifest apparently in suspension beyond the edge of the stage, made its way rather uncertainly around the perimeter and stopped. From it stepped Prince. Neat, very neat.

Within five minutes we had had simulated sex, simulated oral sex and the star had dragged himself across the stage behind the aforementioned Cat, sniffing her in a manner that would have had the most inquisitive dog expressing reservations. But the early lasciviousness detumesced into eroticism and we could devote our energies, liberal attitudes on the ropes but still upright, to trying to keep abreast of a performance that was evolving as though stuck on fast forward, a performance in which everyone shone, from Sheila E, whose thunderous drumming nailed for ever the canard that women cannot drum, to Boni D, whose ravishing gospel voice I would like to have heard more.

Prince himself was everywhere, moving so fast on pencil heels over the staging and amongst the flashing lights and squirts of smoke – imagine a vastly more elaborate and imaginative *Top of the Pops* set under battle conditions – that it was easy to lose sight of him in the mêlée. Seemingly more spontaneous than Michael Jackson earlier this month, Prince, playing Pan to Jackson's Peter Pan, also worked to the split second, now fidgeting with keyboards on a raised, heart-shaped platform, now simulating coitus on a bed that had appeared from nowhere, moving us to introspection with 'Purple Rain' and to a fever with 'Kiss', playing guitar the while with such attack that any, but any, of the world's grim army of strutting guitar fools must have slunk shamefacedly away.

(In passing, one of the pleasures of a show as vital and organised as this lies in knowing that rock legends are not going to get up and jam. They would be trampled to death.)

'I've got a bad attitude tonight, y'all,' claimed Prince, before setting off satyr-like in pursuit of Cat, yet within ten minutes he had us all singing 'God is alive'. The second barely credible show in a fortnight. This standard cannot be maintained.

# Protest Songs

*Observer*, 22 March 1987

ENTHUSIASTIC VAGUENESS passes for scholarship in the twilight world of the disc-jockey. This profundity, which I am having embroidered on a set of tea towels, has come to me during three weeks of researching, writing and recording a four-part series on protest music for Radio 1.

The first problem encountered lay in attempting to discover when protest music could sensibly be said to have started. Sleeve notes to the Peggy Seeger and Ewan MacColl LP *The Angry Muse* were helpful here, chatting warmly of the Anglo-Latin satires of the twelfth century, the works of Wireker and his contemporaries, the Goliadic satire of the twelfth and thirteenth centuries.

In the face of such erudition I beat a retreat to our own century and the seedbed for what most of us understand as protest music, the murderous early years of the American labour movement. Familiar with and much affected by Jack Elliott's recording of a song describing one of the grisliest of many grisly events connected with early attempts at unionisation, Woody Guthrie's '1913 Massacre', I spent several days lost in the trades union section of the BBC record library. I emerged with such gems as Gene Autry's pre-singing-cowpoke tribute to Mother Jones, recorded in 1931, and 'Poor Man, Rich Man', by David

McCarn, a millworker, recorded a year earlier during an especially grim strike.

With the programme written and broadcast, I discovered that these and other recordings my painstaking research had unearthed are freely available on an LP that takes its title from the McCarn song (Rounder Records No. 1026).

This first programme brought in a wealth of criticism. Too scholarly, said one listener – a criticism rarely, if ever, levelled at disc-jockeys – and not scholarly enough, sniffed others. Several correspondents suggested that the programme was little more than a party political broadcast on behalf of the Socialist Party. I countered that by pointing out that we have no Socialist Party in Britain, so imbalance in an election year was not an issue here.

Furthermore, it is in the nature of things that the Right does not, except on its lunatic fringes – programme two included a record in justification of the My Lai massacre – bother to write songs of protest. Records which concern themselves with the uncertainty of pricing in the claret market or the hardship engendered by careless dabbling in copper futures are in pretty short supply.

Programme two brought in further complaints. My favourite came from a listener who objected to the fact that all the records were against something. Hard to avoid this, I pointed out, in a programme of protest songs. The protest instrumental is, as a form, still regrettably in its infancy.

The melancholy conclusion I have reached, after listening to much stirring music, is that ultimately it has little effect beyond reducing the feeling of helplessness experienced by singers and audience alike. Has any government been overthrown by song? Or any wrong righted? I doubt it. Meanwhile we dance on.

# Public Enemy

*Observer*, 29 November 1987

---

'F*** THE NEWSPAPERS,' yells Chuck D, closely echoed by Flavor Flav and most, if not all, of those gathered at the Apollo Theatre, Manchester. By 'newspapers' they mean the *Sun*, the *Star* and the *Mirror*, whose reports on the violence attending – and in the wake of – recent hip-hop performances at Hammersmith have enraged Chuck D and Flavor Flav more than somewhat. When they return to London, Chuck confirms, Public Enemy plan to 'do business' with the authors of those reports. The members of Security Of The First World, flanking the stage and dressed for bush warfare, brandish what must surely be toy machine-pistols appreciatively.

The beat goes on, thick, heavy and hard. This is tricky stuff for us white liberals to come to terms with. For a start, the freedoms demanded by male, American hip-hoppers seem to be for men only. Women feature solely as 'skeezers', whores, bitches or, most unpleasantly, as 'pussy'. Homophobia is endemic and there is the glorification of conspicuous consumption.

When it comes right down to it, I have to accept that the lives lived in America's black ghettos are as remote from mine as those of, say, nomadic Lapps, and hope that the real and justified anger that fires Public Enemy's performances will bring them ultimately to a closer examination of their attitudes to women, gays and paramilitary chic.

Influenced myself by the tabloids, I had been apprehensive about going to the Apollo and had armed myself with a vicious Indian meal. I find a hot blast of onions and garlic most efficacious as a means of repelling unwelcome attentions. Outside the theatre, touts had mingled with a substantial police presence and with local people who had come together, presumably, in the hope of witnessing disorder.

Inside, there was, as far as I could detect, no hint of trouble. The audience was too concerned with blowing whistles or bellowing 'Yo' upon instruction to turn to mayhem. Behind Chuck D and Flavor Flav, their deejay, Terminator X, did important work upon the wheels of steel. 'Everyone who likes the Queen be quiet,' suggested Chuck, and everyone, perhaps taken by surprise, was. It was about the only tranquil moment in an otherwise uproarious evening.

When bill-topper LL Cool J stepped from an enormous radio which had been lowered, to considerable fanfare, on to the stage, the volume increased and the show became noticeably slicker. With DJs Bobcat and Cut Creator working wonders on the wings of a set which looked like the remains of some eighteenth-century folly, LL showed considerable star quality, but the elements of panto apparently inseparable from live hip-hop became irritating. Injunctions to wave our hands in the air, blow our whistles and shout went unheeded, at least by me. Love the records; have misgivings about the live shows.

# Radio 1

## Music Scene 1984

*Guardian*, 20 June 1994

---

IN 1984 Andy Kershaw was travelling back from mainland Europe with Billy Bragg. Andy remembers driving up the ramp out of Dover listening to Peter Powell on Radio 1. Peter was playing a record by Spandau Ballet and, when it was over, he rhapsodised over the Londoners' choice of trousers. That sort of sums up 1984. At the same time that Spandau Ballet were brightening the lives of the nation's trouser fans, Duran Duran were producing videos with exotic locations, healthy young performers and expensive trimmings, which seemed to mock the sick, old and poor on whom Thatcherism had turned its back. In debate at Radio 1, I argued that the smugness and self-satisfaction implicit in these videos were as much political statements as anything conjured up on the left by Billy Bragg, but my sophistries fell on stony ground.

In 1984, George Michael was still 50 per cent – well, 90 per cent – of Wham!, and blending a winning way with a tune with entrepreneurial skill and facial hair research. Madonna was still years away from revealing herself as a care-in-the-community case and making ripping pop records. Boy George (Culture Club), Jimmy Somerville (Bronski Beat) and Holly Johnson

(Frankie Goes to Hollywood) were bringing such colour, commitment and outrage to the charts that it seemed to confess heterosexuality was to demonstrate that you had missed the point.

Ten years later, pop stars barely exist, apart from the factory-fresh creations assembled to plunder the piggy-banks of the prepubescent. East 17 and Bad Boys Inc seem to have few resonances beyond their immediate audiences and, although I feel certain they are, in their own way, little short of delicious, it is hard to imagine they will be troubling us much five years from today. Mind you, they said the same of Elvis.

Away from the pure pop chart, the culture continues to subdivide and mutate enthusiastically into such confusion of genres that it is perfectly possible to find a chart in the pages of, say, *NME*, *Melody Maker* or *Echoes*, apparently related to one's own interests but filled with records you've never heard, made by people you've never heard of, for record labels you never knew existed. Through the efforts of DJs such as Pete Tong, Mark Tonderai, Claire Sturgess, Anne Nightingale and the boy Kershaw, among others, there is now a better chance than ever of hearing the best of these fringe activities on the radio. Recently, Kershaw and I, together with his girlfriend Juliette and my son Thomas, clambered on to a bank below Barregarrow to watch the final practice for this year's Senior TT, to be asked by fellow spectators for details of a Zairian soukouss record they had heard on one of our programmes. This sort of thing happens a lot.

Ten years ago the Radio 1 day started with Mike Read. Mike was followed by Simon Bates, Gary Davies, Steve Wright and Peter Powell. Today, the station – still slightly schizophrenic in its determination to promote new music from whatever source while adhering to a rigid playlist system – retains only the best of these (Steve Wright). The others have moved or have been moved, their antics parodied brutally and accurately by Harry Enfield's Smashey and Nicey.

The greatest pleasure in pop music derives, I believe, from the manner in which its very nature resists scholarship. There have been, needless to say, many attempts at scholarly pop books but most have been either outrageous hagiographies or absurd displays of pomposity. Those that have succeeded have done so because of the excellence of the writing rather than the importance of the subject matter. Pop is a car-boot sale, a parade of trinkets, junk and handicrafts, most worthless, some capable of giving a few moments of pleasure, with a few glorious items made more glorious by their unexpected appearance in this market. Then, in an unpredictable double-bluff, the worthless can, within a few years, take on great worth and the glorious become merely laughable.

I relish this process even when caught out myself. I look forward to discovering in another ten years that much of the music I now admire sounds awful. On the other hand, we still have The Fall, as near-perfect in 1994 as they were in 1984.

# Reading Festival

*Guardian*, 28 August 1995

---

'DANNY BAKER is a twat,' reads Reading 95's top T-shirt. But how can that be, because here I am talking to Danny Baker? We are in the deeply elite free beer compound backstage, envious would-be liggers without the necessary black stamp on their wrists watching us out of the corners of their eyes.

On stage, the Boo Radleys are making Boo Radleyesque noises. Very nice they are, too, but the free beer is nicer so Danny and I chat on. We are talking about football, punk and Courtney Love. The next morning, the *Sunday Mirror* will say that Courtney 'astonished the crowd with her raunchy outfit of stockings and see-through black top'. Of course, the crowd would have been astonished only if Courtney, fast becoming the Queen of Reading, had worn anything else. Love's set with her band, Hole, was, by turns, disconcerting, embarrassing and exciting. The singer, chaotic and brilliant, seems to have become a sin-eater for a whole culture, even for those ghouls who gather in the hope of watching her die. That there is some order, even a measure of contrivance, behind the chaos is demonstrated by the fact that Hole are here at all. It is too often overlooked that this is a first-class band. I hope that Hole return to Reading in '96, preferably with some new songs, and that Courtney survives for years, behaving disgracefully so that her supporters do not need to.

But let's face it, you don't really go to Reading, to any festival, for the music. Not really. How many examples have there been in the longish history of rock festivals of bands turning in that blinding performance that defines them in your head for ever? There have been a few, oddities such as Mungo Jerry at the Hollywood Festival (Hollywood in the West Midlands) and, weirder yet, Police skanking up a storm at Pink Pop, near Maastricht, but, as they say, er . . . that's it.

So let's judge Reading by the extremes. Take Newcastle United-supporting China Drum and Icelandic megabeing, Bjork. Somehow, China Drum have bypassed the Carlsberg stage, an incubator for unknown and virtually unknown quantities and therefore, by and large, the most interesting place to be. China Drum have also bypassed the *Melody Maker* stage, dedicated to those tipped by the music weeklies. Nothing wrong with that. China Drum are on the main stage. Admittedly, they are the first band to play on the first day of the festival but there's a good crowd, not just their mates and their mates' mates either. There's Jonathan, whose first festival this is, with his son, Robin (11), and his daughter, Emma (15). 'These are pretty good, aren't they?' suggests Jonathan, as China Drum pound into another Britpunk stomper. Afterwards, the members of the band are incandescent with glee when interviewed for Radio 1. 'That was great,' they say again and again. It was, too. Even if they never played again, China Drum have played Reading and come away with a good 2–0 away win.

Then there's Bjork, topping the bill on Saturday night. Compère Steve Lamacq has already told us about the stage set. He has seen trees that look as though they are made from wire wool. A technician has tried to describe some sort of super-firework that is going to spell out B*J*O*R*K or something.

I prepared for Bjork by watching a blistering set in the Carlsberg tent from Flinch. The fact that I have had three live radio programmes to do over the weekend has made it difficult to see as many bands as I would have wished and the sheer density of the crowds, something the promoters must address before next year's festival, makes rapid movement between stages impossible.

Earlier, an attempt to get into the *Melody Maker* tent to see Weknowwhereyoulive, the Wonder Stuff less Miles Hunt, had had to be abandoned 15 metres outside because it was so packed. Similarly, a bold initiative after Flinch and before Bjork designed to enable me to catch a few numbers from the Foo Fighters, drummer Dave Grohl's post-Nirvana band, ground to a halt on the back of the crowd some 30 metres from the tent.

The global success of Bjork has been built on her exotic background, the projection of herself as plain weird and a singing style so mannered as to make Kate Bush sound pedestrian by comparison. Finding a clear space some 150 metres from the stage – a clear space in this context usually means a lurking psychotic has been spotted by everyone but you – I watched Bjork pixie about on the horizon. Occasionally, the sound of the Foo Fighters was carried to us on the wind. Despite the magical set and superb sound, Bjork remained stubbornly indigestible, numbers promising to become at least danceable collapsing back into shrill twittering. Around me, people drifted away or announced their intention to stick it out until the fireworks. The fireworks were well worth the wait, too.

# Reading Festival 2

## Here Come the Beers Again

*Sounds*, 10 September 1977

---

'FEAR NOT, dear boy,' they said, 'we'll tow you out again come Monday and no mistake.' And, smiling broadly, they directed me to park the shiny new Peelmobile (well, new to me anyway. One owner, 11,347 miles) in this swamp, this fen, this quagmire, this veritable everglade. (Thanks, Roget. My round, I think.) So for the remainder of the weekend your hero, together with Madge The Simple Village Maiden and Gabs The Appalling Red-Haired Sister-in-Law, had to pass to and from our mobile, chalet-styled home along a narrow plank. I can tell you we scarce dared look down at the turbulent, shark-infested waters beneath us.

*Three days later*

First hitch-hiker: What thought you then, sirrah, of Aerosmith?

Peel (giggling stupidly): People tell me they missed much of their set through being unavoidably detained in the beer tent.

First hitch-hiker (excitedly): Thin Lizzy were fantastic though, weren't they?

Peel (giggling stupidly): People tell me they were terrific, fab, super, epic. Most regrettably I caught only the smallest portion of their set. A prior engagement, you understand.

Second hitch-hiker: But what about Alex Harvey, eh? Sensational?
Peel (weeping softly): I have heard so, I have heard so. But, what with one thing and another, I was unable to view their work. The siren call of the beer tent intervened.

(Peel finds himself wishing – and not for the first time either – that he had picked up the two slim and female hitch-hikers he spotted outside Reading, rather than these eager youths.) Pause for intervention by Editor or Editor's representative. *Are you trying to tell us, Peel, that your so-called overview of Reading Rock '77 is going to consist entirely of gossip gleaned from the coconut-matting floor of the beer tent? Did you see no bands at all?*

Well, Alan (is his name Alan?), the thing about Reading, old cove, is that one spends as much time in socialising, meeting old comrades-in-arms, making (if you're lucky) new ones, and just pottering, as you do grooving to the outside sounds. On the first day – let's call it Friday – I saw Staa Marx, winners of a recent newspaper talent competition (a circumstance which usually condemns a band to oblivion), Salt, Five Hand Reel, Uriah Heep and the Rods. Much of the rest of the time I was nattering of this and of that with Dick Gaughan, he of one of the five or six great voices of our time, and thus missed U-Boat, Kingfish, Lone Star and Golden Earring.

Any band to which falls the tricky task of opening the festival is on a hiding to nothing, and therefore I shall say naught of Staa Marx, save that they look very good and I am eager to hear the tape of their work one of them promised to send me. Salt, who are sometimes represented thus, SALT, were well received. Their old-fashioned blues/boogie music was enlivening and set my feet most furiously to tapping. I felt, and the gels agreed with me, that they are probably at their best in a small and sweaty club somewhere in the Midlands.

Five Hand Reel, who had put together a less folksy set than they would normally perform, were similarly well received. A deftly aimed jig or reel commonly goes down well at Reading, and the Hands wisely played much traditional dance music during their fifty-five minutes on stage. This is not to say that the voice of the aforementioned Gaughan was overlooked. There was much of it, and it sounded as good on stage and in the open air as it does on record. Dick has recorded, by his own reckoning, ten sessions for my radio programmes, yet we had never actually met until Friday. I had expected a morose and suspicious man, one given to squatting for days on bleak mountainsides, preferably in dense fog, and liable to do Englanders like myself to death with a single blow from his mighty claymore. The real man seems modest and peaceable, quick to get his round in, and if my wanderings next week take me as far as Edinburgh then I shall certainly seek him out.

Uriah Heep were hugely efficient. The Rods, on the other hand, seemed somewhat ill at ease, perhaps being sensible that they are poised somewhere between the old wave and the new, between the Reading audience and the *TOTP* audience, and may just slither down into the diminishing gaps between their factions.

As I remarked earlier, I missed the ponderous nonsense of Golden Earring, preferring to spend the late evening enjoying a joke with a gaggle of louts over a stiffish lager. My fair assistants later helped me to bed, or, to be strictly accurate, to sleeping bag, and I remembered nothing more.

Came the dawn and the night was gone, ooooh, what a night! And we left our island home at first light, crossed to the mainland, and prepared to enjoy another day of fast music and faster women. I understand that you want a lot of loose talk about power chords and expanding the language of rock, but your somewhat ineffectual correspondent noticed little until Ultravox appeared. Here is another band which has improved considerably in recent months, and their set was of a very high quality; tight, concise and exciting. (Why is it that I can never think of clever things to say about music? I read other reviewers [*You're a reviewer? – Ed.*] and marvel at their skill with the language, their soaring imagery, their understanding of both the most subtle nuance and the brutal impact of the most thunderous crescendo. Me, I either like 'em or don't like 'em and don't really know why.) So I liked Ultravox whereas I had previously been indifferent. That's what I'm trying to say.

The Little River Band were like one of those extraordinary late-night Australian television dramas. You can spot instantly on whom they have modelled themselves, but somehow the colour values are all wrong and within five minutes of the start you have either switched off and gone to bed or you have popped out for a drink. Mine was a Colt 45. I lay in the van while John Miles played and was impressed with the way he quickly convinced what I had assumed would be a hostile audience. Not at all my cup of tea, but then again neither were Aerosmith, who followed. I have long suspected that the Reading audience is as undiscerning when confronted with notoriety as the cartoon audience which goes barmy for everything featured on that barely believable *Marc* programme on ITV, and this suspicion ripened as Aerosmith thundered away with all the careless spontaneity of a telephone booth.

Aerosmith gave way to Graham Parker And The Rumour, who were just fine, giving me my first genuine thrills of the weekend, with a guitar break in 'Back To Schooldays' boosting my exuberance quotient to unimagined heights. Curse me for a dog, or a fool, or both, but I contrived to miss Thin Lizzy, although I could hear their music and the delight of the crowd over the uproar of those unable to battle their way into the liggers' enclosure in front of the stage. My travelling companions expressed a guarded satisfaction with the band, although several confirmed Lizzyites I spoke to subsequently were of the belief that the band has reached that dangerous point in their career at which being well-drilled replaces being genuinely exciting.

Of the three days of Reading it had been the Sunday for which I was waiting most keenly. Here were Motors, Blue, Racing Cars, and Frankie Miller, plus several others I was not at all averse to seeing. We breakfasted well before gambolling out into the mud to see the day's openers, Quartz. They did extremely well, delighting that section of the congregation anxious to be incited to put its hands together and to hear heavy metal songs about demons. The ageing smartass such as myself may regard this sort of thing as being more than a hint passé, but we forget that there is still a huge audience for such things, an audience not prepared to endure the more contemporary whimsies of such as the Electric Chairs, whom they were later to pelt with large quantities of the only ammunition Mother Nature had vouchsafed to them, mud.

I had rather fondly imagined that The Motors, favourites of mine, would get a pretty enthusiastic reception for their high energy but yet tuneful work. They didn't, and I still don't understand why. I can't, being an opinionated bugger, accept that I am wrong about them and they are actually no good. Perhaps they were on too early. Anyway, I spent the next coupla hours commiserating with them, and thus missed Tiger and the 'A' level rock of The Enid. Blue too were but indifferently received, their music being perhaps too delicate for the essentially – and rightly – rowdy atmosphere of the event. No one seemed to notice the shadowy figure in beret and violet raincoat who slipped on to play pianoforte on their final number.

You're going to hate me even more for this, but I was away gathering football results as Racing Cars played. Distant cheers told me that they were doing well. I was too bush trying to compose an ode to Liverpool FC – I gave up when I could find no rhyme for Dalglish – to witness the rout of Wayne County and his cohorts.

Sunday night at Reading is beginning to take on several ritual aspects. There are the disgraceful chants which insist that I, as compère, am miraculously transformed into a key part of the female anatomy; there is the drunken bellowing of 'Hey Jude' and 'Nellie The Elephant'; the menacing growl that follows the first bars of whichever fearful record by the Brotherhood Of Man is being currently flogged to death on Radio 1. Bands fortunate enough to play on Sunday night are assured of a good reception at the very least, and after the luckless Electric Chairs had flitted away, Hawkwind, the Doobie Brothers, Frankie Miller and finally Alex Harvey and ensemble were generously received. I had being looking forward to seeing Frankie Miller above all things, and despite being engaged in conversation throughout his fifty minutes, and despite being sufficiently intoxicated to ask Frankie's manager if he was going to sing 'Brickyard Blues' while he was actually singing it, I was not disappointed. Here is the logical successor to Rod Stewart, and his singing of 'Jealous Guy' cut the master rather seriously. And, according to The Appalling Red-Haired Sister-In-Law, he has really nice green eyes.

As you may have deduced from the above, I heard less music but enjoyed myself more this year than at any previous Reading Festival. I probably drank too much and ate too much, certainly talked too much. We Peels tend to rabbit on when happy, and the lads parked alongside us, the mystery pianist, a couple of Motors, Dick Gaughan, Phil Lynott and a whole host of greater and lesser unfortunates suffered as a result of my contentment. If you've got this far you can count yourself as my latest victim. I had hoped to complete this account without a single 'I' or 'me'. As you can see, I have failed ignominiously.

But there's always next year.

# Record Shops

*Punch*, 16 January 1980

---

THERE WILL BE those amongst you – I feel certain of this – who will remember the dear, dead days of the mid-1950s; days before commercial television, when to be mentioned in an Honours List was not the mark of a bounder, and when professional footballers, still cheerfully unaware of the special relevance to their calling of ill-health among parrots, arrived at League matches by public transport.

For the record collector this was the Golden Age of the listening booth, and many and fair were the afternoons the pubescent Peel spent huddled in the basement of Wildings of Shrewsbury catching up on the latest teenbeat sounds, only to saunter out, when I had worked my way through the catalogue, with nothing more than a (free) copy of the Top Twenty. Not that I never bought records, but as a schoolboy with an unwholesome craving for pikelets, I relied heavily on an interesting range of gullible relatives for keeping my fledgling collection fit and bursting with rude spirits whilst my own resources were diverted into the school shop.

I recollect being given 78s by my grandfather at a very early age, and the old gentleman's judgement was sound enough for him to have bought for me 'The Thing' by Phil Harris, and I had, over a period roughly bounded by the end of World War II and the end of rationing, hammered into smithereens my father's first-class collection of pre-war records, many of them works which now merit peculiar reverence when discussed on Alan Dell's splendid *The Dance Band Days* on Radio 2, but it wasn't until the Christmas of 1952 that I first purchased a record with my own money and without outside assistance.

Despite the well-established popularity of the gramophone, there were, even then, very few shops which traded exclusively in records and a Radio 1 colleague assures me that as a horribly pimple-strewn youth he worked behind what I expect was called 'the record bar' of a shop the principal business of which was the sale and repair of fishing tackle. He also remembers a middle-aged man popping in to buy a little something by Mantovani for the wife, and asking whether they would be so good as to put a tune by that Nat 'King' Cole on the other side.

For my first hesitant steps into the record market I had selected Crane's of Liverpool (and, almost certainly, Birkenhead and, I shouldn't be surprised, New Brighton) as recipients of my custom. Crane's dealt primarily in furniture and musical instruments – perhaps they still do, in which case they have my blessing – and I had to pick my way through several acres of gleaming pianos forte (this is almost certainly not the approved plural, but I rather care for the cut of it) before coming to the record counter, itself an insignificant subdivision of the sheet music department. Sweating hideously over the 4/3d clutched in my hand, I begged, in thrilling tones, for a copy of the newly released 'Blue Tango'. Without a word the assistant took my money and handed me a Columbia 78 in which Ray Martin and His Concert Orchestra had recorded the number of my choice, and I scurried away, rat-like, into the street. I have that copy of 'Blue Tango' still. Would you like me to hum it for you?

Since this nerve-racking beginning I have bought far, far too many records, even having two rather artfully chosen collections stolen in their entirety, and still spend between £25 and £30 a week on the blamed things. Much of this money is squandered in minuscule specialist shops physically and spiritually at several light years removed from the glossy houses where cowed customers line up to buy whatever is currently the subject of extensive television advertising. In these latter the only words ever heard are 'Sorry, dear! If it's not in the charts . . . next!'

By contrast the sordid basement, stalls, and tiny shops where I buy records are important centres of social activity, hotbeds of gossip and intrigue, where a savage delight is taken in not stocking any record which may be had of W. H. Smith's. Despite their deceptively casual, even anarchic, air, these dwarf businesses call for highly specialised shopping techniques involving subtle colourations of speech, politics, movement and costume. Accents should be industrial, preferably Cockney, although a Belfast accent goes down well, and the politics are substantially to the Left of the Labour Party and include – and quite rightly – implacable opposition to the ultra Right-wing mobs which have visited extreme and unprovoked violence on a considerable number of concerts and dances during the past two years.

Tapping the toe in a carefree manner to the music being played in the shop is a poor idea, studied boredom being de rigueur. (May I here digress into an exposition of Peel's Law, which states that records which sound marvellous when heard over an in-store stereo system sound flat and tedious when replayed in the privacy of the home.) A barely perceptible movement of the shoulders is just fine should you wish some outward expression of the rhythm fettered in your soul, and a riggish flutter of the hips is dandy if you have the figure. I have not.

As far as costume goes, there are only two rules. (1) Don't wear flared jeans. (2) Don't wear flared jeans.

One of the many attractions of the smaller shops is the pleasing anonymity which is allowed to cloak a Radio 1 disc-jockey picking listlessly through the racks. I say this without vanity – it is a fact of life that BBC disc-jockeys are, in varying degrees, recognisable folk. I'm not insinuating that you personally have a Radio 1 calendar hanging up somewhere, of course, but there are a lot that do.

This minor-league notoriety only becomes a problem when I go shopping in one of the bigger stores, because I only do this when I am seeking some long-lost nugget from my youth or some especial horror for use with the overpaid John Peel Roadshow (which is me and two boxes of records). Assuming, almost certainly without foundation, that other customers may be peeking at my selections to learn what is, you know, really hip this week, I go through agonies as I strive to keep the Alma Cogan or George Lewis and His New Orleans Jazz Band albums hidden behind yet another copy of the latest Clash album. I have five of these at home. If you write me a nice letter I might send one of them to you.

# Cliff Richard

*Observer*, 10 November 1985

---

MUCH HAS been written about Cliff Richard's apparent agelessness and from the back of the Hammersmith Odeon he certainly looked lithe and fit last week. Twin runways sloped from the back of the stage to the front and, during curiously uninvolving instrumental passages, Cliff was given to indulging in Bowie-esque mimings and struttings on them. The suspicion gnawed that some of this posturing was perceived as interpretative dancing.

Between numbers Cliff spoke to us. His own songs, he insisted, dimpling modestly, were no good. The well-mannered audience clucked sympathetically. Much of his music, he went on, he classed as 'rockspell'. The world changed fast, he wanted us to know, but Jesus was to be trusted at all times.

A rockspell song or two later Cliff told us that, 'as a Christian, I don't mind singing about the nuclear age'. We could stop worrying about the bomb, he explained, because Jesus would look after us. The temptation to yell out in disagreement was considerable, but soon quelled by the prospect of a severe buffeting from what looked like a congregation of gym teachers.

This platitudinous piffle led into a song probably called 'Living Under The Gun', during which the deft use of smoke and flashing lights recaptured something of the romance of life in the trenches. 'The world could die tonight,' crooned Cliff, firing imaginary guns from his hips and, goodness me, lasers flashed tracers right into the audience.

No sooner had we recovered from this excitement than Cliff was off again. In some country he was too coy to name – could it have been South Africa? – he had come to understand the value of personal freedom. His solution to the lack of this commodity was to 'Leave this town and play some rock 'n' roll.' Do the Mandelas know this?

During the interval I left and retired to licensed premises. There can be little doubt that Cliff has tussled with sin and by and large won. But he does seem to be having a spot of bother with the sin of pride.

# Road Rage

*Radio Times*, 29 January – 4 February 1994

---

A COUPLE OF weeks ago I was driving towards Mill Hill Circus in north London at the end of the rush hour. There are four lanes into the roundabout system from the south, two taking traffic round to the right, two leading straight ahead towards Watford and the A1 or left into Mill Hill Broadway. It often happens that drivers unfamiliar with the system get into one of the right-hand lanes, discover at the last moment that they have made a mistake, and swerve left. It happened again this time and I was forced to take evasive action to avoid becoming the middle car in a scrap sandwich. A four-track to my left appeared to brake to give me room and I waved my gratitude.

This, it turned out, was a serious mistake. The driver of the four-track brought his vehicle up on my inside at some speed and, realising that mine is a left-hand-drive car, deliberately drove at me. Again I was forced to take evasive action and, looking left, saw my assailant, a large man with a manner that indicated that violence was not unfamiliar to him, screaming abuse at me. He was under the impression that I had made a 'V' sign at him and, with much use of the one word that still seems to be *verboten* on radio and television, warned me that providing traffic flow conditions were favourable at the top of the hill, he was going to leave his vehicle and kill me. Frankly, I believed him.

His last question before he disappeared in the direction of Watford (the lights at Northway Circus were, thank God, green) was the one I mused over all the way here.

'I've got an effing kiddie here, you ****,' he said. 'What do you think he thinks about your effing "V" signs?'

I have no doubt that, had the lights at Northway been red, I would be reporting to you from intensive care or not reporting to you from the morgue. Since this incident, I have watched programmes on the rise in violence with heightened interest. What signals, I wonder, are given to a child by a man who is prepared to wreck his own car and who threatens the life of a stranger because he believes a 'V' sign has been made at him?

I was musing on this later that night as we watched *Living with the Enemy* on BBC1. Sheila and I paid particular attention to the home videos and the spitting, shrieking rows between mothers and their daughters and, comparing them with what goes on in our house and in those of our friends, we wondered whether anyone at all leads the warm, funny lives depicted by advertisers and sitcom producers. Is the best you can hope for a sort of *Roseanne*-style standoff? (How will Roseanne, back on Friday, be with Darlene at college, by the way? Darlene is the character I admire above all others on television and I shall miss her.)

The rows we have in our house are relatively mild and tend to be the result of parental frustration at the infinite variety the young persons can bring to the important work of winding each other up, although I have lost my temper with each of them at one time or another, usually when circumstances switch me into last-straw mode. You know the sort of thing; rotting food in bedrooms, Doc Martens left on dark stairs, homework abandoned in favour of soaps, records removed and not returned, all the standard stuff. But there were enough familiar moments in the first *Living with the Enemy* to make it uneasy viewing.

From *Enemy* we switched over to the *Anglia News and Weather* and from that slid involuntarily into *Network First*'s documentary about women who have killed their partners. This set me to thinking again about my would-be assailant.

There has been so much necessary stuff written recently about the banality of evil, that to refer to systematic brutality, whether in the killing fields of Bosnia, or the living rooms of Britain, in those terms is to risk accusations of cliché, but the most memorable aspect of *Women Who Kill* was the firm, undramatic manner in which the victim women catalogued the often grotesque brutality they routinely endured. They grew hesitant and ill-at-ease only when they told of the moments in which they had put an end to their suffering. Well, I say 'an end', but the women had much still to suffer from the anomalies and absurdities of the justice system.

As *Network First* ended, Sheila, William and I sat, not saying much, counting off the remaining twenty minutes of William's childhood. At midnight, Tom, having learned it specially, came down from his bedroom to play Happy Birthday on his guitar for his older brother. Without that we might not have slept so well.

# Roadshows

*Disc*, 14 April 1973

IF ANY ONE of our 243,776,431 readers is named Richard and plans to marry a smart woman with thin blonde hair and the merest rudiments of a chin, then I have a warning for him.

I travelled with her and her equally frightful friend on a tube from Liverpool Street to Oxford Street and overheard her discussing the changes she plans to make to her consort once the gate has slammed shut behind him. She spoke about him as though he was a flat rather than a human being. Get out, Richard, while you can. The preceding has been a public service announcement.

I had my first two gigs of the year last weekend and they were, as always, an unqualified success (advt). When I reached the Hull Arts Centre with my caravan of go-go women, performing bears, trampoline experts, unfrocked bishops and professional footballers, the streets were already crammed with well-wishers who hoped to catch a glimpse of me before I adjourned to my dressing-room to address myself to the task of selecting the most desirable of the virgins provided for my sport. I handed out well over a thousand signed photographs and 'John Peel is Terrifically Exciting' T-shirts.

Once inside the Arts Centre you could sense the tension and excitement that thrilled through the expectant crowd which had been jammed into every available space in the tiny theatre for well over two weeks. Tickets had exchanged hands for as much as £325.

I can't keep this sham up a moment longer – and I suspect that there were those among you who suspected that I was making it all up anyway. The reality was that I arrived early, walked the wind-swept streets of Hull alone, played football against derelict houses for an hour and provided a wearisome hour of unpopular records until Jo'burg Hawk came on and saved the day.

As the regular DJ was to memorably observe the next day when we were in a wind-swept and rain-sodden Cleethorpes, many people approached him afterwards and told him that he was 'much better than that John Peel'. Just the sort of thing I love to hear.

Also in Cleethorpes I was approached by a drunk who told me that he found my programmes infinitely dull, that he wondered why the BBC employed me and that none of his friends thought I was any good either. He then expressed surprise when I asked him whether he was that unpleasant to every stranger he met.

There are, fortunately, always nice people wherever you go and they often appear from the most unlikely places. The manager of the Dreamland in Cleethorpes and his lady were about the only people who spoke to me all evening, other than to request records. They took me to a pub afterwards and we sat and talked until the small hours.

In my own perverse way, I enjoy doing gigs, wish I got more of them, and understand that it's entirely my own fault that I don't. People pay their money to see something slightly larger than life and when they don't get it they grow peevish.

For my next live appearance (scheduled for early 1983) I intend to staple crows on to my cheeks, wear only éclairs woven into a form-fitting garment and to arrive on stage after free-falling from Concorde. I plan to stand and shout at the audience for an hour, striking the smaller members of that body with silver mallets, and finish my 'set' by burning the building to the ground.

This 'live' show will be recorded and will be released as a quintuple LP called *Hull '83*. I hope you, my wonderful, wonderful fans, who have put me where I am today, will buy this small token of my esteem for you.

# Roadshows 2

*Sounds*, 1 March 1975

---

WHAT WAS IT Kasimir had said in the Tilsit Biergarten? Eric smiled to himself as he remembered that crisp November morning. It seemed so long since he had seen Debbie, so long since he had crushed her in his arms, rained kisses on her eager, upturned face.

Dare he risk lighting a cigarette? The border was only 150 metres away, bathed in an unnatural orange glow by the flares. At irregular intervals a searchlight beam stabbed into the night. Sinking lower into his hiding-place, Eric reached into his pocket for the crush-proof pack of hand-rolled Vendome Thins.

Placing one between his frost-rimmed lips he checked for the hundredth time the Belgian snub-nosed Rijtborg-Visper 35-8.2 in its hiding-place between his ears. Yanto Vonx would have laughed to see him now, crouching in this alien ditch only 150 metres from freedom. Ruefully he rubbed the bruise the Yugoslav had left on his cheekbone and realised, with a start, that he hadn't shaved for two days.

Speaking of bruises, the John Peel Road Show is a mass of them.

Last weekend – it seems so long ago now – I was at Tiffany's with Jack the Lad, Snafu and Medicine Head. Jack the Lad had been on, playing a crisp little set (sounds like a lettuce, doesn't it?) crammed with amusing songs, jigs and reels to set the feet a dancin', dancin' all your cares away, and an atmosphere of

Geordified rowdiness. The stage had revolved, transporting me in the twinkling of an eye from my position to the extreme right of the stage to a less notable position amongst the debris backstage.

Now if there's one thing that makes your Uncle John ever-so cross it's having roadies test their band's equipment when he's playing records. Seems silly, I expect, but even if you're just running a second-rate disco you like to do it right, and having folks shouting 'one, two, one, two, two, two, two' over the top of your carefully considered chatter doesn't come within my personal notion of what is 'right'.

I don't recall any 'one, two'-ing from Snafu's roadies, but the band's gear did seem to be suffering from a severe and apparently incurable attack of the feedbacks, and after five or six minutes of intermittent howling I resolved to suggest to whoever was responsible that the orchestra would be somewhat put out if I elected to test the disco gear during their set – would he, I wonder, do me the courtesy of saving the uproar for later?

Thus resolved, I set off at a crisp trot for the front half of the stage, motoring smoothly as I did into a handily-placed heating duct. The result was about quarter of an hour of having very little idea of who I was, what I was doing or why.

Now, several days later, I have a most impressive bruise turning seductively septic on my vaulting brow. To those who have inquired from whence it came I have told a heart-warming story of driving away, with my bare knuckles, a gang of thugs who were molesting a young girl. My audiences have been unfailingly sceptical, believing, I'm afraid, that the injury was more likely to be the result of an unsuccessful molestation on my part.

As the Noel Edmonds Breakfast Show gets under way on the Roberts, I must needs peer into my diary to see where the new and self-flagellant John Peel Road Show (Two Balloons And A Torch) will be annoying perfectly respectable people this coming weekend. Well, on Friday 28, our nationwide tour brings us to Hull University.

On Saturday 29, I shall spend the afternoon at Anfield watching Liverpool humbling Chelsea before pushing the Road Show handcart over to Bolton Institute of Technology. I bet they're all keyed up in Bolton, eh? On Sunday 2 March, we come, bathed in green limes, to Mr George's in Coventry.

I don't fully understand why I'm getting all these gigs – at times I feel like the subject of some complex nationwide sociological project. How long can he endure having people spitting 'Black Sabbath' and 'Status Quo' into his face? How many nights must he sleep in lay-bys before he starts to imagine he's Napoleon and declares his kitchen a Free State? How many times can he hump that sodding box of LPs up five flights of specially greased stairs before he sustains a rupture? We Peels are made of pretty stern stuff, researchers, so don't imagine our breakdown is just around the corner.

Oh Noel! He's just spoiled my morning by playing that Guys and Dolls single. Why is it that all media people, especially those responsible for finding guests for TV 'Variety' (a classic misnomer) programmes and for selecting records to be played on the radio, wake up every morning hoping to hear new New Seekers? Don't we, the long-suffering public, deserve better? Evidently not.

I forgot to mention it earlier, but for only £2 in cash, cheque or money order, you can have a beautiful, full colour reproduction of the poorly painted 'still-life' of a peach and a bottle of gripe-water that hangs in the studio when I do my oh-so-wonderful programmes. And if you don't know what on earth I'm talking about then you're not the watchers of the benighted *Crossroads* that I take you to be.

Yesterday I cycled 28 miles for charity. I want you to know that I too am full of good works. I also have a bruise the size of a moorhen's nest on that part of me where the saddle partially fitted. If you hurry I'll let you see it before it fades.

# Rock's in Trouble

*The Listener*, 22 November 1973

---

IN 1954 Elvis Presley sprang loud, greasy and menacing from the cab of his Mack truck to do battle with Daughters of the American Revolution and parent-teacher groups all over the States. With him Rock was born, after a lengthy gestation, and spotty youths like myself at last had a hero – or a choice of heroes, some of whom even committed the then ultimate sin of being aggressively black – who we could be sure would horrify and scandalise our parents. Almost at once, tiresome pundits were forecasting that Rock was only a passing menace and that we'd soon be fox-trotting again to the Frankie Laines, Johnny Rays, Perry Comos, Max Bygraveses and Ruby Murrays who had previously dominated the charts. They were gloriously wrong, as wrong as they were a year or so later when they assured us that the half-baked lounge-lizard calypso of Harry Belafonte would rid the air of all these Fats Dominos, Little Richards, Jerry Lee Lewises (and this monster's thirteen-year-old wife). In succeeding years and ever more shrilly, they advised us that the mambo, the hootenanny and sundry other horrors were going to flush away Rock for all time. They even resorted to the ancient cry – a cry, incidentally, that can still be heard in the corridors of Radio 1 – that 'the big bands are coming back'. They were always wrong because the kids always wanted something that parents would abominate.

Every time my father complained about 'that bloody awful row' or hinted that Gene Vincent couldn't sing, the more devoted I became to the music, revelling in its adult-proof mysteries.

In the early sixties, Rock did fall ill for a couple of years. The businessmen deprived us of the crude, glorious energy of rock 'n' roll and fed us instead vapid treatments of tired old songs by debonair and wholesome bores like Bobby Vinton and Bobby Vee. Records of that era now have a certain camp value, but it was a depressing period to live through. Fortunately, the Beatles came to our rescue, bringing in their train the Stones. Once again, the young could be sure that their parents would be deeply mortified by and distrustful of the long-haired, unwashed and doubtless over-sexed monsters whose pictures were disfiguring their daughters' bedroom walls. Once again, all the traditional parental values were being challenged – and they have undergone even greater shocks since, with the fashion in the late sixties for hallucinogenics and now the apparent sexual ambivalence of many rock stars.

What can a beleaguered Mum or Dad do now? Well, sadly, he or she can regard the charts and take heart. For Max Bygraves and Perry Como are back and are as bland as ever. David Cassidy (the modern Eddie Fisher?) is there, and so are legions of Osmonds, cuddly and uninspired as Doris Day ever was. We have the Carpenters too, a duo with the energy and sexuality of a broken leg, the Simon Park Orchestra, and a lad, one Michael Ward, who is the toast of the Darby and Joan outings with a number entitled 'Let There Be Peace On Earth'. Besides these aberrations, there are many pre-pubescent formula rock records by the likes of Gary Glitter, Mud and Suzi Quatro. The charts haven't looked so dull in twelve years. The problem is that there's no one in there that parents can loathe, and when the Mums and Dads are crouching to peer at *Top of the Pops* with all the enthusiasm of their off-spring, it's a sure sign that Rock's in trouble. Hell, they even like David Bowie and he's alleged to be multisexual at the very least. Is that the sound of Ambrose I hear borne on the wind?

# Rockpool

*Independent on Sunday*, 17 June 1990

---

RECENTLY I HAVE contrived, after three or four years of unseemly begging, to be placed on the mailing list of a New York-based operation called Rockpool, which circulates monthly reviews, playlists and general information, along with records, to its members. Since the importers Shigaku ran into financial trouble and my friend Ian moved to Nottingham, Rockpool has been my only source of American music. I mention this because there is something radically different from our own about the American independent record sector. I am talking vast generalisations, of course, but the best of the Yank records sound as though they really needed to be made; as though, indeed, there was no process that could prevent them from being made. This makes them works of art. Too much British music is made in the hope that the major labels, circling overhead, will swoop and devour.

There is something cantankerous about America's independent labels, even a sort of frontier spirit, which seems more than several thousand sea miles from the cowed, slightly submissive attitudes of most British labels. You want an example? Well, there is Jim Gibson's Noiseville, operating from PO Box 124, Yonkers, New York 109710, on a budget so tight that $200 owed by a British distributor can threaten the whole operation.

Releasing editions limited to 500, 700 or a thousand, Jim runs Noiseville because he loves the wild, confrontational music of Bootbeast, Abu Nidal, Bench, Unholy Swill and The Goatmen, no more expecting to enrich himself in the process than whoever puts on gigs at the Sea Cadet Hall, Cambridge.

Leaving Harlow, having heard Levellers 5's soundcheck with one and a half of their top-quality songs, I was thinking about what singer/guitarist John Donaldson had quoted from Mark E. Smith of The Fall: something to the effect that all he wanted was to be able to provide the members of the band with a decent wage. Somehow that struck a chord.

# Shaving

*Sounds*, 1 July 1978

---

THERE ARE TIMES, reptiles, when your Uncle John, bless him, wonders, as he goes merrily about his daily round, whether rock stars could get away with being as much adrift from the spirit of the music as some prominent ex-players seem to be from the essence of football.

Last night, having witnessed, with rising gorges, the luckless Dutch being again hustled out of the world title, this time by a combination of a standard of refereeing that would draw gasps of astonishment in our local youth club league and the gross misconduct of Argentinian players who realised that they could do no wrong, those of us grouped disconsolate around the TV leapt as a person to our feet, baying with fury, when the awful, sanctimonious Jimmy Hill trotted out something to the effect that 'football isn't only about winners, it's about losers too', and deplored the fact that the Dutch hadn't appeared to take part in the post-game ritual. If I'd been one of the Dutch players I wouldn't have left the dressing room at anything less than gunpoint and even if I'd been told I'd never play football again if I didn't.

One slight consolation from the passing of the World Cup is that we will no longer have to suffer this Charlton or that giving us blinding insights into the game of the order of 'a goal at this stage of the game would make a real difference'. In about thirty years attending football matches I have yet to see a goal that didn't make a real difference. But enough of football, eh, and let's have some more about me.

Those of you who could be bothered to ferret about among the small ads last week, and thus locate my column (this, by the way, is the bitchy bit) sandwiched between details of Carpenters bootlegs and heart-rending pleas from Swedish boys (18) who want to know everything there is to know about Paul Anka, may have noticed – in fact, your attention would have been drawn to it – that there were two snaps attached of Peel in the shaven or beardless mode. Why was this? I hear you whimper. I shall tell you.

Ten years ago, when I was but a boy, straight of back, clean-limbed and sexually active, I thought, on a whim, to grow me a beard. Accordingly, the family doctor, when consulted on the matter, recommended I stopped shaving and, sure enough, over a period of months a beard grew. Well, what with one thing and another – you know how it is, I'm sure – the beard stayed, and silly old Peel grew silly and old beneath his luxurious pile.

From time to time I felt a bit of an urge to have a look at the face lurking beneath. The face that had bewitched an entire generation of young women, young women delicate and as soft as swansdown, swansdown as – wait a minute, this is getting out of control. Let's do the decent thing and start this sentence again . . .

From time to time I felt a bit of an urge to have a look at the face lurking beneath, and would trim the growth with a pair of ivory-handled shears my dear old grandpapa brought back from one of his expeditions to discover the source of the Orinoco (hands up anyone who thinks I'm playing for time here). Whenever I did this I was disturbed to see my mother peering at me through the pruning. Now my mother is a charming woman, a delightful companion and as deft a hand with a chilled lager as I've seen outside Lords, but I don't want to look like her, so the beard has stayed in situ (look, if you don't give a toss about my beard – and who can blame you for that – why don't you move on to another part of the paper? There must be something else here you like).

Last year, when we bearded chaps were drawing fire from Jimmy Pursey and the likes – you probably didn't realise that some of us grow the farting things for aesthetic reasons and to prevent outbreaks of hysteria among the impressionable young – one found oneself under a lot of pressure to shave and be damned. Last week, my own curiosity whipped up to and beyond bursting point, I bowed to this pressure and, having first drunk about two-thirds of a bottle of Libyan Liebfraumilch, and with the assistance of Madge the Simple Village Maiden, removed the tangled undergrowth.

To my horror, the strapping, handsome youth who had vanished from sight ten years ago had been supplanted by some raddled old sod with a long pointed nose and a 'chin' which extended from his lower lip to his breastbone in one loathsome swell of fat. This latter monstrosity is, you can bet your life on it, vanishing behind my hair again as fast as modern agricultural technology can grow the stuff. I may shave it off again to bemuse my great-grandchildren, but not before. Nossir!

# Sick in Trains

*Disc*, 1970–1

THIS HAS BEEN a fun-packed week. Last week, you will doubtless recall, your Byronic hero (me, you fool) was immersed in soapy water with a young woman. This resulted in the long run in my being unable to introduce the contestants on *Friday Night Is Boogie Night* and in the Pig ordering me to bed at eight o'clock.

At three o'clock on Friday morning, after two hours' sleep, I was woken by sharp and persistent pains in my rather lovely stomach. These ultimately forced me into The Smallest Room where I spent many a fretful hour being sick and doing other things too frightful to mention. By nine o'clock I was well enough to raise myself up on one elbow and whimper rather pathetically for a breakfast preparation we call Pigsli.

Having downed this and having arrayed myself in all of my costliest finery I walked carefully out to Friday (a Land Rover) and readied myself for the drive into London.

Once, twice, three times I turned the key and nothing happened. Half an hour later I was still turning the key although alternating the turns with a lot of hopeful peering into the engine. I am a man whose mechanical knowledge just encompasses the less complex kinds of toothbrush.

Finally, abandoning hope, I phoned for a taxi and headed for Stowmarket station and caught a train which, perusal of a timetable assured me, would coincide, at Ipswich, with another train which would speed me into London. I had not noticed that the second train only ran on Saturdays.

Ipswich station is not the sort of place I would immediately recommend to those suffering from stomach distress and a nagging urge to vomit colourfully and in large quantities. By the time the train arrived I was convinced that it was only a matter of time before every one of my internal organs discharged itself through one or other of my bodily orifices. (I hope I'm not distressing you.)

One of the more diverting things about travel on this particular line is that you either have the entire train to yourself or share it with enough people to fill Wembley Stadium several times over. On this occasion I was not alone and found myself standing in the corridor between a Welshman, who thought it was scandalous, and the biggest truck-driver I've ever seen.

When my illness became apparent and I was disappearing into the bog for the sixth or seventh time, the truck-driver became very helpful and solicitous. Each time I reappeared he'd clear a way for me and stand me by the window. Each time I disappeared he'd wait a minute and then shout through the door to find out whether I was OK or whether my worst fears had been realised. He was so friendly and kind that I felt a lot better when we finally reached Liverpool Street. It is reassuring to meet people like that.

The last time I was ill on a train – the same line curiously, but about five years ago – the passengers completely ignored me even when I finally collapsed on the floor (I was on my way to hospital from Radio London with blood poisoning). When we reached Liverpool Street on that occasion someone called the gendarmerie because they felt I must be maddened with drugs.

Carrying a box of records on rush-hour tubes is never a heap of fun but the legendary courage of the Peels drove me on and I came at last to the BBC. By this time I was well enough to search about for a tender message from the Sue Hook I mentioned a week or so ago but as none had come I had a sleep instead.

When I woke I felt a great deal worse and finally conceded that I would have to let the fearful B*b H*rr*s do the programme. The business of getting back to Liverpool Street and the train journey home were so unpleasant that I really don't like to think about them. All the way I consoled myself with a vision of a quiet night in front of the television with Pig and Pig's Sister ministering to my every need. When I got home it was a very different reality that faced me. A harsh and brutal Pig drove me up to bed and wouldn't even let me read. I hope you all feel very sorry for me.

As this is a music paper I'd better mention some musicians. Let's see now. I did get a postcard from Robert Wyatt and his lady this week – and the famed trumpeter John Walters has been scaring the wildlife around these parts with some of his own amusing variations on well-known and much-loved tunes.

A lady called Elizabeth has written a friendly letter from the village. She gave neither her address nor her last name so I'm scrutinising every face I see to see whether or not it looks like an Elizabeth.

We're listening to the Bob Weir LP again – we can't listen to *Never a Dull Moment* all the time – and the kitten, fast becoming a cat, is asleep on the table beside me. Woggle (a dog) is tugging at my trousers because she wants me to go and play football with her – and if you think I'm making that up I have photographs to prove it.

I get quite a few letters from *Disc* readers these days – just friendly letters about the sort of small things with which I fill this column – and that really is nice. There are lots of good people about and if you're one of them, thank you.

# The Smiths

## Morrissey Unmarred

*Observer*, 2 August 1987

---

IT IS WITH regret, runs the press release, that 'The Smiths have to announce the departure of Johnny Marr from the group. However, Rough Trade would like to confirm that other guitarists are being considered to replace him.'

When word of The Smiths began to seep out of Manchester in 1982, the most arresting aspect of the band, unheard, was its name. Apparently banal, except to citizens named Smith, it implied nothing, gave no statement of intent.

When Rough Trade issued the debut single 'Hand In Glove', a record which, almost uniquely for the period, bore no clear evidence of influence from other musicians, it was evident that this was a band to be enjoyed.

On the second of the two occasions on which I have met Morrissey, singer and lyricist with The Smiths, we were both customers of the motorway service area south of Newcastle on the A1. This was, I was later assured, a watering hole the angular singer particularly relished. This, even if totally untrue, would be very Morrissey. Since he first drifted into public view it has been clear that here is a man determined to live his musical and personal lives at his own tempo and in his own way, keeping the foolishness of the rock industry at the other end of a generously proportioned barge-pole.

Johnny Marr, on the other hand, and according to a report in this week's *New Musical Express*, is eager to bring his own considerable and individual skills as a guitarist to collaborations with such inherently un-Smiths-like entertainers as Keith Richard and Bobby Womack. An unscheduled flight to America to record with Talking Heads is rumoured to have caused this final rift.

Critics have long insisted that The Smiths are talent-free miserablists. This strikes me as complete nonsense. The first time I met Morrissey was on a record review programme in which he delivered himself of carefully worked-out and often outrageous assessments with a delight in language that was almost nineteenth-century.

More than one Morrissey lyric has caused my laughter to tinkle among the teacups. His ability to indicate a whole way of life by briefly highlighting a darkened corner of that life is matched only by his skill at delivering his lyrics in a manner that leaves the listener with no choice but to consider seriously what is being sung. These are rare gifts in popular music.

My initial reaction to the news that Morrissey and Marr can no longer bring themselves to work together was one of distress. Subsequent musing has persuaded me that both men will, if they can remain their own masters, continue to surprise and amuse.

# Sock Syndrome

*Radio Times*, 12–18 August 1995

---

SOMEONE THE Aardvark Goes Quantity Surveying. Oh, come on. Was it Eric? Eric the Aardvark? I was at a Promenade concert, my first in twenty years, and these musings had sprung unbidden and without a backstage pass into the dressing room of my mind. Below me, Sir Peter Maxwell Davies was conducting the European première of his own 'The Beltane Fire'. I was not enjoying it. Nicely polished shoes he's wearing, though, my mind suggested, in a positive sort of way.

I was at the Royal Albert Hall as the guest of BBC radio supremo Liz Forgan and had, in my turn, invited brother Alan to join me. The invitation had specified lounge suits and as I don't own such a thing, I had taken my nice blue shirt from the cupboard and put on my clean black jeans. Alan had loaned me a jacket but it was a pretty swampy night, so I wasn't wearing it. Rather, I had it flung casually over a shoulder in the manner, I liked to think, of a French film star – albeit a French film star with reservations about the wisdom of renewed nuclear testing.

Our little group had gathered before the concert in the General Scott Room. Alan and I were the first to arrive. As the other guests checked in, we noted how suave and cultured they all were and how prone they were to inviting each other to pop in and see them next time they were in Oxford. Alan is in television and can switch on suave and cultured whenever he likes. As the room filled, I could catch him saying, 'Ah yes, I worked with Humphrey in the seventies,' before going on to tell some tale that had his audience guffawing appreciatively.

I had found Radio 2's droll and apparently unflappable Ken Bruce and was clinging to him desperately. Despite the heat, Ken was not sweating. Neither, it seemed, was anyone else. I already looked as though I had but recently been plucked from wind-tossed seas by Air Sea Rescue units scrambled from Wattisham.

By the time we were chivvied into our boxes for the show, I had become aware that my shirt was suffering from what I think of as Sock Syndrome. At some stage in the washing cycle, it had snuggled unwisely against a sock that has spent some weeks fermenting on a bedroom floor, with the result that, as I heated the shirt from within, a rich smell of unwashed sock started to pollute the area in which I stood. As hysteria welled in my ample bosom, I grew hotter and the reek worsened. In addition, the shirt, taking on a life of its own, started to knot itself about my body as though anxious to become as one with my flesh. Meanwhile, as we sat down, everyone else looked as crisp and fresh as your proverbial lettuces.

Altogether unable to concentrate on the music, my mind set off in several different directions simultaneously. Why did the first conductor look like that Hewetson boy I had disliked at school? Why was my younger brother so much more socially adept than me? Why had no one invited *me* to pop in next time I was in Oxford? Even, in an unreconstructed moment, would the cellist with the bleached hair and interestingly sinewy shoulders ever turn around?

By the interval, I had relaxed a little, reconciled to the certainty that any conversation my fellow guests would have in the rest of 1995 would, after the usual observations on how lovely Oxford was at the time of the year, turn to the little man who was so damp and smelly at Liz's Prom.

The second half of the programme was given over to Rachmaninov's Piano Concerto No. 3. Actually, according to the programme, decent folk now spell Rachmaninov thus: Rakhmaninov. This sort of thing happens quite a bit in serious artistic circles, I believe. It gives the impression of cultural movement, of innovation and rigour, where, in truth, none exists. I have decided to take a further step and go for Rakkmaninov. Are you with me?

My spirits soared with both the appearance and playing of soloist Grigory Sokolov. Grigory looked as though he could be the bass player with the third-best blues band in Preston, but he played like a god. During the portions of the work in which the orchestra falls silent and checks its fingernails, Grigory flew, crouched Muppet-like over the piano, humming and buzzing along with the music as he played. I wanted to clap between movements, to leap to my feet and yell, 'Go, Grigor, go!', perhaps even to dance. Instead, I just carried on sweating – but happily now.

'He's just a thumper,' someone suggested later. My kind of thumper, pal.

# Sound City

## John Peel Records a Hectic and Demo-packed Week in Sound City

*Guardian*, 28 April 1995

---

Monday
Arrive Bristol after ill-advised detour in holiday traffic to Broadstone, Dorset, to take tea with rural technolords, Distorted Waves of Ohm. First person spotted in luxury hotel is Leena, my control at Finland's Radio Mafia, for whom I have been doing programmes for eight years. Leave for Anson Rooms of Bristol University in Finn-packed taxi. Tonight's Evening Session popsters are Menswear (better than cynical, hard-bitten DJ expected) and Supergrass (great records, slightly rockist live). Finn-less walk to the Malaap venue in intermittent light showers. The band I had hoped to see (Secret Shine) finish as I arrive, amusing music journalist with whom I was to drink beer and talk nonsense announces he is exhausted and is going home to bed. Hear Blueboy before solitary walk to hotel. Cornered by member of Menswear plus female acquaintance who could sulk for Britain.

The demo tape count: 9 so far.

Tuesday
At luxury breakfast (tinned mushrooms, hash browns) everyone

telling me how good gigs I missed (Elevate/Bugg/Orbital) were. Spent afternoon lying in Castle Park outside tent covering Radio 1 stage. Dogs, children and demo tapes proliferate. Some ripping bands (Mondeyspider/Cake) but have to leave before Doyenne, in order to walk miles, mainly uphill, to meet Planet Record moguls Richard and James. Driven back downhill to docks to join thirty or so mariners for trip around colourful, heritage waterways. Planet and Teenagers In Trouble, who have won my heart by attempting to remake whole of first Woodstock LP without benefit of musical ability, play as we sail, to evident displeasure of evening strollers who come within earshot. Hearty walk back to hotel. Can't figure out local taxi systems. Meet Claire Sturgess, Radio 1's Queen of Rock. Persuade her and World Service duo to walk with me to The Louisiana for Planet showcase featuring rare performances by Flying Saucer Attack, Crescent and Movietone. Pub packed, veteran DJ exhausted, fearing undignified collapse, but sustained by sizzling if untidy presentation on stage. Separated from Queen of Rock so solo walk back to hotel. Top Radio 1 management figures in bar and getting their rounds in. Favour them with unsought advice whilst peering over their shoulders for members of Elastica, who have been playing the Anson Rooms. Don't see any.

24 demo tapes.

Wednesday

Had planned motor trip to Newport, Gwent, and legendary venue, TJs. Feeling unaccountably glum. Cancel trip. Plan instead to return to The Louisiana for Crush/Heads/Tribute To Nothing. Don't do that either. Decide at last minute to make way to Kings Arms, Blackboy Hill, for Baby Harp Seal. Can't find it. Retire to room 928 to contemplate 31 demo tapes.

Thursday

Sort records acquired during week. Arrange demo tapes in

meaningless piles on dressing-table. Worry about what I am to say at Watershed Cinema event billed as 'A Conversation With John Peel'. Arrive ninety minutes early and stew in café with coffee, sandwich. Cinema packed. Started nervously but customers benign and ninety minutes pass without incident. Meet up with Mike Hawkes and Alison Howe from my own programme in Malaap where supposed to encounter amusing music journalist (see Monday). No sign of him and music not to our taste. Search for food ends in pizza hovel.

46 demo tapes.

Friday

Collect daughter, her boyfriend and, by coincidence, Queen of Rock from station. To high-grade second-hand record shop and buy *It's Trad, Dad!* soundtrack for pounds 10. Chair *Guardian* pop quiz. To Anson Rooms to broadcast from van in car park, playing teen terrific records and introducing the Bluetones, Pulp and Dread Zone on stage. This involves starting disc then flying up four flights of stairs to effect introduction. Arrive breathless each time but manage amusing announcements which daughter later tells me are completely inaudible. All bands perform like Liverpool in 2–0 win over United. Smug DJ reckons has brought best music of week to Radio 1 audience. To bar to receive congratulations of record industry together with poncey foreign beers but no one says a word. To bed.

53 demo tapes.

Saturday

More record shopping. Live programme from Bristol's version of Broadcasting House. Confirm status as worst interviewer on planet but eloquent guests save day. Strike out for home after dropping off Norwegian DJ with poorly chosen aftershave. Bowl along M4 with grin on weather-beaten face.

67 demo tapes safely in boot.

# Viv Stanshall

*Guardian*, 11 March 1995

---

KEN GARNER'S *In Session Tonight* lists nineteen sessions involving Viv Stanshall recorded for Radio 1 programmes I introduced, seven in which Viv was a member of the Bonzo Dog Doo Dah Band and twelve under his own name recorded between 1970 and 1991. Numbers included such arcana as 'The Craig Torso Show', an early assault on DJ culture, and 'The Bride Stripped Bare By The Bachelors', a joke quite lost on me when I broadcast it in autumn 1968.

With the sessions under his own name, usually supervised by John Walters, one of too few people strong enough to bring Viv's wayward abilities under necessary control, Stanshall introduced listeners to the grotesques who inhabited the mythical Rawlinson End, in a sequence of sketches under such titles as 'Giant Whelks At Rawlinson End', 'Gooseflesh Steps' and 'Cackling Gas'. In these complex, magical fantasies, Viv was assisted by musicians such as Andy Roberts, Zoot Money, Barry Dransfield, John Kirkpatrick, Dave Swarbrick and Danny Thompson. The months spent in preparation for the recording sessions, the last-minute delays and the increasingly bizarre excuses offered by Viv for non-completion would have driven most producers to abuse, recrimination and cancellation. It is a tribute to Walters, not a man celebrated for his patience, that he persevered and to Viv that this perseverance was worthwhile. The resultant pieces swung wildly from nonsense songs, through dense poesy to deliciously cruel flights of fancy. I can never see hang-gliders without recalling the understandable pleasure Sir Henry took in blasting them out of the sky over Rawlinson End.

I admired Viv's wit, imagination and lunatic sangfroid so much there were times when I would have wished to be him. It has been for twenty-five years a cause for regret that Viv's wayward and self-destructive behaviour meant that so little of what he had to offer took tangible form. He was, on his day, the funniest man in Britain.

Our paths crossed rarely – I was rather alarmed by the fully unfettered Stanshall – but I have an enduring memory of the Bonzos playing at Hatfield Polytechnic. Viv stood centre stage in what looked like pyjamas and swinging about his head a length of flexible tubing that had been adapted as a form of trumpet. About him musicians flailed, cavorted, laughed and pranced, before him students laughed, danced or stood in bewilderment. Amid the chaos and uproar, Viv stood cool and relaxed, his air that of a foreign office mandarin charged with bringing the latest news of Imperial folly to a particularly brutish outpost.

He was a great man and it has been our good fortune to catch some of the echoes of this greatness. I think Viv would have enjoyed knowing that a hundred years to the day before his own terrible death, the death was recorded of Assyriologist, soldier, consul, discoverer of the Persian cuneiform vowel system and more, much more, Sir Henry Rawlinson. To the day, mind you.

# Viv Stanshall 2

## What's So Funny?

*The Listener*, 7 July 1977

DID YOU KNOW that the most popular boy's name in Britain in 1900 was William? And that, in 1975, William had slipped to a bad forty-first? Two years ago, the most highly regarded name for a new male Briton was Stephen, and it is some measure of the troubled times in which we live that the name Jason had sprung from nowhere to pin down the coveted number fifteen position. Have we, as a nation, lost all our traditional values? My father, a decent and generous man, would never allow the owners of Jaguar cars, when visiting the family home, to park their motors in our drive; and I like to think he would have refused admission to any little Jasons seeking to enter the house, on whatever pretext.

I was toying with these and related matters as I prepared myself physically and spiritually for the keeping of this 'Diary', which I did by driving halfway across Suffolk, with my wife and son William to take beer in the gardens of the White Horse, a fashionable house on the A1120 outside Yoxford. The solid information that fuelled my musings comes from the *Sunday Times* colour supplement for 22 May. I find that whatever non-musical information is retained in my head these days comes from the colour supplements to our two great Sunday newspapers, and that both of these are generally stuffed with the type of semi-precious information a simple man can lower gently into conversation, thus giving the impression that he is a chap of some erudition. Last Sunday, for example, the *Sunday Times* printed a short piece about celebrated Scottish droll Billy Connolly, presumably, at one time in his life, another William. In the course of this piece, it was revealed that Billy has suggested that a part of the reason for his success, which has been and is considerable, lies in the fact that 'British comedians on the whole are so bad.' And I think he is right – and I speak with the clear authority of a man who, at the age of eight, sang 'Fuzzy Wuzzy Was A Bear' on stage at the Liverpool Empire with Frankie Howerd. If you call hoarsely for a further qualification, then consider this – I once had a brief part in an episode of *The Goodies*, possibly the least amusing 'comedy' show ever witnessed by humankind. I only got the part, I should hastily add, because Ed Stewart, friend to the nation's Jasons, was unable to take the job. Even then, all I was called upon to do was to pretend to be Jimmy Savile, OBE. A fairly humbling experience, if you like, being an Ed Stewart surrogate affecting to be Jimmy Savile introducing a cod *Top of the Pops* on *The Goodies*. A man who has sunk that low has known pain. But I digress.

Where, then, are the new British comedians, the successors to the grand old tradition of Tony Hancock and all the others? Most of the professional Uncle Jolly-Boys we witness on television are either, if they are old-established 'comics', extravagantly unfunny, or, if of the new breed, well on the way to becoming extravagantly unfunny. You want examples? Well, in Category A, Morecambe and Wise, whose best work in several years is the current television commercial for Texaco. And in Category B, just about any of the men who emerged from Johnny Hamp's influential *The Comedians*.

These latter, predominantly northern English and, at first,

sharp, observant, accurate, and very, very funny, were men who knew that although breaking wind and impersonating Edward Heath could be amusing in their place, there was a lot more to comic life than that. Since *The Comedians*, many of the best of them have been siphoned off into, and shamefully emasculated by, that class of television bromide known, with a callous disregard for accuracy which must surely border on the criminal, as 'variety'. You know, Rolf Harris, Harry Secombe, Mike Yarwood, Cilla Black – that sort of variety.

Overlord of BBC TV variety is Bill Cotton Jnr, surely never a William, and a man with his finger so firmly on the public pulse (some might say throat) that he was able to say, with marvellous smugness, when asked recently whether the Sex Pistols would be seen on BBC television, that 'people do not wish to see such things'. (I had hoped to reach the end of this piece without mentioning the Sex Pistols. Regard them in this context, if you will, as merely symptomatic of a far greater malaise.) Obviously, a lot of people *do* want to see the Sex Pistols, but they are equally obviously not the right people; so what Cotton meant, I fear, was 'we do not want people to see the Sex Pistols'. But what evidence is there that people are keen to see, for example, Dukes and Lee, Hope and Keen, Miguel Brown, and David Hamilton – as they were given the opportunity to do a Saturday or so ago? Billy Connolly's reflection on British comedians has neatly crystallised a vague disquiet I felt the other day while listening to Radio 4's *Boy Meets Girl*, described in *Radio Times* as 'a confection of recorded humour on the eternal encounter and its consequences'. Which is exactly what it was. I noticed then that all the really good stuff came from the 1950s and 1960s, with the exception of the odd spot of John Cleese.

In passing, it is worth remarking that Americans on the prowl for something the valuable Francis Howerd would characterise as 'titterable' have an even worse time of it than we do in Britain. Since the early 1960s, when such as Bob Newhart, Jonathan

Winters, Shelley Berman, Lenny Bruce and Mort Sahl were working so well that some of them saw LPs of their live performances nipping up to the very top of the US charts, Americans have had to make do with the juvenile 'hipness' of the dreadful Cheech and Chong – and worse – for their chuckles.

How sad it is, then, that in Jubilee year (I know that has nothing to do with it, but I was determined to work it in somewhere – keep your eyes skinned for 'darling of the Centre Court' next) the man I consider to be the funniest Englander I have heard works only intermittently, and then only – and I apologise for what must seem like blatant self-advertisement – on my Radio 1 programmes. I speak of Vivian Stanshall, formerly of the Bonzo Dog Band. Viv is only now recovering from a prolonged battle with ill-health, a battle that made him erratic and very unreliable, but his most recent work for producer John Walters and myself has been marvellous. Viv currently bases his work around 'Rawlinson End', a sort of warped and dreadful Blandings Castle, and the outrageous prejudices and leisure activities of its hellish residents. Viv tells those fearful tales, which are punctuated with perverse little songs of his own devising, in accents redolent of a Britain-that-never-was, falling somewhere between Wodehouse and Capt. W. E. Johns. To divorce a single line from the 'Rawlinson End' context is to invite bellows of 'that is not at all funny', but the image of one frightful Stanshall bounder, who wears sunglasses with frames in the shape of Ford Cortinas, is one which stays with me constantly and can send me off into fits of giggles under almost any circumstance.

Now that I have two hours a night on Radio 1, rather than one, we are able to give more time to the shocking fantasies of Viv Stanshall, and his most recent work for us will be repeated on 12 July. I hope you will have time to listen for him – particularly if you are a record company. The man should be recorded now and often.

# Sub Pop

*Observer*, 29 January 1989

---

THERE IS a paper to be written about the geography of rock. Why Liverpool? Why Manchester? Why not Birmingham? And why Seattle? And why, while we are at it, now? A 1981 compilation christened *Seattle Syndrome* dropped no hint that anything even remotely interesting would ever happen in Washington's largest city. Yet in the past eighteen months the name of Seattle has popped up whenever two or three have gathered together to speak of non-chart pop – and as often as these folk have said 'Seattle' they have said 'Sub Pop'.

The first release on the Sub Pop label, a compilation called *Sub Pop 100*, featured such grade A noisemakers as Sonic Youth, Scratch Acid and Steve Albini. Anyone prepared to dismiss this as an opportunistic one-off might have felt differently had they noted the message on the spine of the *Sub Pop 100* sleeve. 'The new thing: the big thing; the God thing: a mighty multinational entertainment conglomerate based in the Pacific Northwest,' it ran.

With the exception of Steve Fisk, now a producer and recording engineer at Velvetono Studios in Ellensburg, Washington, none of the *Sub Pop 100* artists had any discernible connection with Seattle; but since 1986 the label has focused on Washington state bands, with releases from Green River, Soundgarden, Blood

Circus, Swallow, Nirvana, Tad and Mudhoney. Common to these records has been the most identifiable house – as opposed to House – sound since early Tamla, a thick, angry, pressure-cooker guitar/bass/drums/vocals turbulence. Mark Arm of Mudhoney, whose severely limited edition single, 'Touch Me, I'm Sick'/'Sweet Young Thing Ain't Sweet No More', provided a right marker for 1988's releases, has spoken of the desire to produce 'a kind of tidal wave of noise'.

With genuine limited editions, a coloured vinyl and a unique promotional style – 'Destroyed the morals of a generation' (Green River) and 'So what if they smoke pot? This sounds like . . . a head-on train wreck' (Blood Circus) – Sub Pop could not be ignored. Bruce Pavitt, the label's founder, had realised the importance of Sonic Youth (who have recorded a version of 'Touch Me, I'm Sick' for the label), Big Black and the Butthole Surfers to those on both sides of the Atlantic enraged by the drift of the post-Joy Division generation of bands towards cuteness and had set out to do something about it.

The latest something is *Sub Pop 200*, a boxed set of EPs featuring not only the label's own bands – Mudhoney contributes a barely credible version of Bette Midler's 'The Rose' – but groups from other area labels such as K (Beat Happening, Girl Trouble), Empty (Cat Butt) and Popllama (the Fastbacks, the Walkabouts). It is going to take something special to stop *Sub Pop 200* being the set of recordings by which others are judged for some time to come.

Mudhoney and Soundgarden (now signed to a major label) tour this year. The distant roar is the sound of queues forming. The God thing is coming.

# Summer of Love

*Radio Times*, 7–13 August 1999

---

ROBIN'S BOY, Jim, wants to know what the difference is between the sixties and the nineties. Let me tell you what I think, Jim. Now, I'm not one of those souls who sits back with a faraway look in his eyes, blows imaginary smoke out of the corner of his mouth and murmurs, 'Ah, the sixties.' So although I had a good time in the sixties, mine is not, I hope, an assessment tainted by nostalgia or false memory syndrome.

Essentially, Jim, in the sixties the general assumption was that, by and large, what with one thing and another, life was going to get better and better for everyone. Oh, there'd be many a slip along the way but things would improve, not just for hippies, not just for Australasians, North Americans and Europeans, but for people in remote villages in Sierra Leone and on the coast of Bangladesh as well. Quite what this 'better' entailed no one seemed entirely certain, and there was similar uncertainty as to where all this theoretical improvement would lead us. But I suppose those of us who thought of ourselves as hippies imagined that somewhere down the line there would be a moral revolution to match the scientific one. This hasn't, as you will have spotted, happened yet, despite bold talk of Ethical Foreign Policies, and I don't believe it is going to happen on New Year's Eve, to the accompaniment of meteorites, earthquakes, tidal waves and Voices On High, either.

In 1999, Jim, the general assumption seems to be that things are going to get steadily worse, not just for the villagers in Sierra Leone and Bangladesh, but you and me and Becks and Posh and just about everyone, and that all you can do is hope to retard the process a little.

Jim – and Robin – are among those interviewed for *My Generation* and I feel I know them well enough to address them personally because I did the commentary on the film. There are disadvantages to doing this sort of thing. Principal among these is having to discuss a programme with people who believe you have written the script and have a burning interest in the contents. Sometimes I do, sometimes I don't. I usually explain to my tormentors that doing these voice-overs is a form of prostitution except that, happily, I'm not called upon to get naked, that you sit in a darkened room, usually, appropriately enough, in Soho, and when a green light flashes you say, your voice dripping with desire, concern, excitement or whatever seems appropriate, either: 'It was then that Debbie decided it was time to return to Dulwich,' or: 'The Errol-Johnston PN26 had the revolutionary reversible fly-wheel.' You do this without the slightest notion who Debbie is or why she left Dulwich in the first place or without, of course, the remotest idea as to the benefits of a reversible fly-wheel. You then pocket a disappointing cheque and rush home to feed your family and face middle-aged men at parties who insist on telling you, at wearisome length, about long summer days spent with their mates Snap and Lofty in the marshalling yards of some northern town with a football team clinging doggedly to a place in the Second Division and how they once saw the Dumfries Flyer and one of the last 243s in the same week. In 1962, wasn't it, Marjorie?

I felt some fellow feeling for Robin and Jim because Robin, now a lawyer, went to an Isle of Wight festival in the late sixties (so did I) and Jim went to Glastonbury this year (so did I). I didn't, to be honest, go to the same Isle of Wight festival as Robin

– the famous Hendrix and French anarchists one – but I did go to the one the year before, with the Jefferson Airplane. All I can remember was that it was bitterly cold and, for reasons I promise not to go through again, I had no socks.

Everyone who writes about festivals goes on about naked people, and there certainly are naked people in this film. This is more to do with the fact that if you take your clothes off everyone in the county with access to a camera will be within viewfinder range in a trice than with any tendency to wanton nudity. I've been to loads of festivals and have seen depressing little nakedness – and I have been on the lookout for it, let me tell you. You'll meet some interesting people in *My Generation*. Unless you're a festival-goer yourself, you'll be astonished at the amount of dope-smoking that goes on and, I'm afraid, the amount of piffle that is talked. Well-intentioned naïvety, grumbles Robin, speaking of something altogether different, and that neatly sums it up. Speaking personally, I'd sooner spend my hard-won leisure time in the company of the naïve but well-intentioned than the sophisticated and malevolent and I appreciate the sense of community that still, despite rampant commercialism, pervades events such as Glastonbury. Some people may be out to charge you ridiculous prices for a glass of red wine or to steal your children's tents, it's true, but it is worse in the real world. I can understand those who believe the real world is best avoided at all costs.

# Today Programme

*Radio Times*, 22–28 April 2000

---

BE STILL my beating heart, be still. I've just been on Radio 4's *Today* programme talking about drugs. Politicians apparently love being on *Today* – I don't. Not really. I'm flattered to be asked, of course, but I sleep poorly the night before and don't usually deliver the goods. This morning I started from my bed at about 4.30, only to lie back down again and spend an hour running through the useful, realistic, thought-provoking things I was going to say. People's lives were going to be altered for the better by them, I thought. Then I fell asleep again. In the event, I was neither useful, realistic nor thought-provoking. I've only once contrived to be even mildly coherent on *Today* and that was towards the end of last year. I can't remember what it was I was talking about on that occasion but I spoke clearly and slowly and made sense. This morning I gabbled. Gabbled and waffled. As you might, I suppose, on certain illegal substances.

I don't really bother with drugs, unless you count red wine and vegetable korma as drugs, but I do have opinions on them, as most people do. My not bothering probably has more to do with laziness than anything else, but if I was told I had six months to live, I think I'd try them all, just to see what the fuss was about. There are so many conflicting opinions, so many scare stories, so many eager campaigners for and against that it's difficult to know the truth any more. What the children do is their business, but I suggested to each of them in turn that the greatest danger lies in the necessary contact with people whose interests are served by encouraging you to experiment further and that you couldn't really be certain what it was you were buying anyway. Daddy was, I told them, with a reassuring smile, a bit of a smoker himself back in the sixties. Each of them looked at me as though I had suggested that they drink their own urine every night upon retiring.

Unfortunately, this morning I hadn't progressed beyond my breezy introductory remarks and was about to say, 'But seriously, Jim,' when it was all over. Drug tsar Keith Hellawell, who followed, contemptuously brushed my contribution aside. He was probably right to do so. I'll have a vegetable korma tonight to salvage my wounded pride.

# Tommy

*Sounds*, 5 April 1975

PICTURE THE SCENE in London's gay and giddy West End. It is the night of the world premiere – make that World Premiere – of Ken Russell's film of *Tommy*. Outside the cinema are hordes of curiosity seekers, television technicians, fans, pickpockets and police informers, all pushing and shoving in the hope of catching a glimpse of the stars.

A straining thin blue line of security men is struggling to hold the throng behind the ropes, and away from the luxurious deep pile of the flame red carpet that runs from the very edge of the pavement to disappear into the freshly painted foyer of the cinema.

From time to time, in a hundred grimy restaurants in the streets around about, swarthy waiters without a word of English turn to one another, carefully avoiding the eyes of patrons as they do, to shrug eloquent shoulders as the echoes of the crowd, bellowing with excitement as some fabled name swans, smiling fixedly, into the theatre, penetrate their seedy world. Gangs of Moniques and Susettes, French models from Clapham, wriggle resignedly beneath sweating businessmen from Leicester, and dream of what might have been.

Another roar from the crowd, Rod Stewart has been spotted! Already they have moaned their ecstasy as Roger Daltrey, Mick Jagger, Bob Harris, Gary Glitter and countless other stars have passed their view into the brilliantly lit cinema. There is nothing quite like a worl … sorry, World Premiere.

Here is Elton John, there is Oliver Reed. See them laugh, heads thrown back uncaring, as they enjoy a joke.

But stay! The crowd falls silent as round the corner, the lights of London town dancing and sparkling on the brilliant, polished, blue paint, comes, slowly, a 1926 Hispano-Suiza. 'Iris, it must be John Peel,' murmurs a pasty-faced secretary, 'he's here, he's here,' as she slides to the floor at her friend's feet in a swoon. Iris, standing on tiptoe straining against the crowd to see the great man pass, doesn't even witness her companion's distress.

Anoushka, my blonde Russian émigrée chauffeuse, brings the Hispano gliding to a stop against the pavement. Across the road I catch a glimpse of Feathers, the under-chauffeur, who has spent a whole week with oils and lacquers of his own devising, bringing a special and magical sheen to the car. I acknowledge him with a curt nod, the crowd sobs with excitement as I do.

On the other side Pete Townshend, Princess Grace and several lesser members of the Royal Family pass unnoticed – the masses have only eyes for me. How they envy me my elegance, the correctness of my dress, the straightness of my carriage. Maidens sigh longingly as their eyes fall to the reinforced trouser front that an over-generous nature has compelled me to wear.

I stop and shake the trembling hand of a pensioner and present her with a 42-page document which explains how she can claim supplementary benefits. The crowd applauds spontaneously at my generous gesture. How important it is to keep some sort of contact with the people, eh?

As I pass through the doors, brushing aside the stars who press forward to address me, the crowd begins to break up, to start the trek back to their modest homes in Hampstead, Hammersmith and Hendon, content in the knowledge that tonight they have been in the presence of the truly great.

Oh, alright then. I wasn't invited.

# Too Decrepit to Walk

*Radio Times*, 2–8 March 2002

I DON'T LIKE, even now, to talk about my time in the army. The nation's enemies are everywhere, as you know. Walls have ears – and there's the Official Secrets Act to worry about as well. So, generally speaking, mum's the word. However, yesterday – and I don't believe I'm betraying the country's interests in telling you this – it became necessary for me to deploy some of my military skills in a rather special way.

It all started with a phone call Sheila made to her sister, Gabrielle, who lives in the village. They were planning a walk. When Sheila got off the phone, I asked if I might join them. 'Blow away the cobwebs, don'cha know,' I harrumphed. Sheila looked dubious. 'Well, we are going quite a long way,' she said. 'And we don't dawdle, either.'

The inference was plain. I was judged too decrepit, too out of condition, to walk the sort of distances she and Gabrielle were contemplating, and I'm sure you can imagine my chagrin when, some twenty minutes later, the two women yomped off up the road together, taking with them Bernard and Nellie, our dogs.

'I'll get on with some work, then,' I shouted, but they weren't listening.

There was a cold wind blowing when, five minutes after their departure, I set off up the hill behind them. I already knew that at the top of the hill Sheila and Gabrielle would turn down the muddy track, but nevertheless I checked the ground for the broken twigs and wellington and paw prints that would indicate that they had recently passed this way. With nostrils flaring attractively, I hurried past the Suffolk County Council sign that announces, with admirable economy of style, 'No Right for Horses. Public Footpath Only', making pretty good time along a path turned into a quagmire by, well, horses. After half a mile, I caught sight of the two women crossing a field. Crouching in the ruins of an old cottage, I observed them until they were out of sight before continuing.

A few hundred yards further on, I met a mother and daughter on horseback. The path was narrow at this point but they made it clear that it was my responsibility to get out of their way. I tugged my forelock as a gesture of submission. As my forelock is located these days at the back of my head, they may have thought my gesture threatening. Certainly, as I stepped aside into a cluster of brambles, they deemed it wiser not to run the risk of thanking me.

For another hour I continued to stalk my prey, now standing motionless, until they had rounded a corner, now loping at a low crouch alongside a hedgerow to avoid detection. Some two hundred yards from the house I caught up with the sisters as they stood talking in the road.

Sheila wanted to know what I was doing. 'I am only obliged to furnish you with my name, rank and serial number,' I countered. Under further questioning I broke and admitted that I had been stalking her and Gabrielle. 'Oh, for God's sake,' she said.

You can sleep easier in your beds tonight knowing that I, at least, am in a state of military preparedness.

# Top of the Pops

*Observer*, 1 September 1985

---

THE NOTE the commissionaire passed on warned that Alix and Melanie were at the gate for Mr Peel. Alix and Melanie (their real names) had been at the gate for Mr Peel since mid-morning. They were also at the gate for Miss Long. I passed them, pausing only to explain that it was unlikely that I would be able to get them into the studio, when I arrived at the Television Centre at 1.30. I passed them again when I hiked in twenty minutes later from a temporary BBC car park in, I think, the Lake District.

By 2.00 I was in the studio, attempting to adjust to life in an atmosphere consisting mainly of vapours produced by dry ice machines and receiving instruction from producer Michael Hurll. These consist of details of assignments – 'John, you've got Bananarama and Madness. Janice, you've got Bryan Ferry' – and details about timings. I was once given eleven seconds in which to interview Debbie Harry.

I have a theory, untainted by research, that no one listens to the DJs on *Top of the Pops*, that they are peering over our shoulders, as it were, to catch a first glimpse of the star for whom they have a hankering. Unfortunately these stars do seem to listen, and Paul Hardcastle, who is playing some sort of keyboard device with D Train, wants to know what I meant when I said eight weeks ago that his record '19', which concerns itself with the average age of American Vietnam war victims, had come nineteen years too late. He seems to accept my explanation because he later consults Janice and me over the salads available in the canteen.

At some moment lost in the mists of pre-history I must have said something beastly about the Thompson Twins, because they – or their representatives – once required that I should not introduce them on *Top of the Pops*. They are on the programme again and when I passed the band's Alannah Currie on the steps down from Stage B, she fixes me with a basilisk stare. Later she invites Janice round for a meal.

At 2.15 we record the Chart Rundown, for which we are out of vision, getting it right first time. Word has it that other DJs have needed as many as fourteen takes. We feel pretty pleased with ourselves. At 2.30 we rehearse our introductions to the Top Forty Breakers and the Top Ten Videos. This is a more complicated business altogether, as we are in shot one moment, out the next, then back in again. This requires fine timing, with the sheets of yellow paper, on which we have scribbled our prettily worked ad libs, hidden behind our backs until the camera lights go out.

At 3.45 there is a complete run-through of the programme. Our drolleries produce no reaction whatever from the dancers, cheerleaders and technicians. On the second run-through I introduce the word 'bottom' into my opening remarks and this goes down quite well. By 5.20 rehearsals are over and I retreat to my dressing room for an hour's sleep. Later I escort Janice Long to the bar where Dan Hartman, whose record 'I Can Dream About You' is at number sixteen, joins us. I don't much care for 'I Can Dream About You,' but Dan turns out to be a most affable chap. We have a serious talk about, amongst other things, the proliferation of toll-roads in Connecticut.

At 7.20 I report to make-up. I always dread this, believing that the make-up artists would rather be chastised with scorpions than apply the dense layer of powder and paint that serve to make me look three or four months younger.

In the studio the cheerleaders, the real heroes of *Top of the Pops*, are whipping the crowd into a frenzy. Tonight's DJs, they bawl, are Janice Long (cheers) and John Peel (silence). We clamber up to our positions on stage, the floor manager signals that there is one minute to go and we run through our opening remarks yet again. Janice checks that there is no lettuce on my front teeth. Another *Top of the Pops* is about to take to the air.

# TT Races

*BIKE* magazine, September 1992

---

THE HEDGE outside R. Bradley's Garden City Store at the bottom of Bray Hill has grown several inches since last year and I have to stand on tiptoes to see over it. A couple of bus-passers amble over to ask this good-looking stranger whether you get a better view from this side of the road. I ask them, my voice rich with condescension, if they have been to the TT before. It seems this is their seventeenth year. It is my second.

And I wouldn't have come last year if Andy Kershaw had not arranged transport and accommodation and I had not been too frightened to say 'Er, well, Andy, the truth is I don't think I can make it actually.' Afterwards I bored everyone rigid with it, explaining how astonished I was that thousands of people from all over Europe could come together in a confined space, consume absurd quantities of strong drink and not kick each other's heads in, and how wonderful it was, after the prices, predictability and showbiz bullshit of Grand Prix motor-racing, to see some real white-knuckle stuff again.

So here I am outside the Garden City Store, hoping that Mr Bradley will invite me to stand on his garden path, the better to see the contestants bearing down the hill.

This year I left things so late that the only way I could get onto the island was by air. Now I'm a chap who doesn't much care for flying. Devil-may-care as you like at zero altitude but above that, well, unease sets in by the vanload. Thus it was that when Andy collected me from Ronaldsway after ninety minutes on a plane with propellers – and, in my view, with not nearly enough engines – on which I had been issued with boiled sweets, I was not in perfect race trim.

Those who know me will affirm my dedication to the notion of my body as a temple fit for my God, but it came as a shock to discover that those muscles which are normally assigned the job of preventing me from breaking wind in public were being given responsibility for clutching on to what I am horrified to learn is called the 'buddy pad' of Andy's Harley.

However, after bowling along in the sunshine for a spell my spirits were lifted, to be sent soaring by the sight of a portly man painstakingly brushing gravel on to the road a few miles south of Douglas, presumably in the hope of unhorsing someone who had strayed somewhat from the racing line. At this I laughed out loud and called roughly for beer.

Andy and his pals had taken a flat for the duration but I had persuaded your editor – and my new boss – to insert me into a first-class hotel. If you have paid for this copy of the magazine you have also paid for my room. Thank you. If it is any consolation, I didn't have the breakfasts.

I think it is time I laid before you my own motorcycling credentials. During my military service – and I can't tell you more without compromising national security – I ran a James 150 which I bought when my stepsister wanted her Lambretta back. Er . . . that's it.

Mind you, I do have a rather nice leather jacket. Unfortunately the elaborate motif on the back advertises a Mexican restaurant rather than some experimental braking system but, as I always say, you can't win them all, eh?

I was also planning to use this year's TT as a sort of consumer's guide, hoping to see somewhere amongst the thousands of machines scattered about the island the instrument nature intended for me. The problem is, as I reflected walking along the Prom from the luxury hotel to the Kershaw encampment in the poor part of town, you never seem to see two bikes the same and, as I further concluded, with the sea breezes tugging playfully at my hair, too many seem to have been decorated by the design team that brought us the shell suit. But something was bound to turn up. At the Kershaws' the talk was of Status Quo. Some felt we should, as fearless journalists, attend their concert but I pointed out that if we quelled the merry popping of corks for a minute, we would almost certainly have borne to us on the wind the distant strumming of guitars and choruses of 'Down, down, deeper 'n' down'. Calmed by this, we trekked back to my hotel in search of famous motorcyclists.

Something else that impressed me last year was that as a member of the public and without the muscle of *BIKE* behind me, I still managed to have my photograph taken with Robert Dunlop and Trevor Nation. Try getting a photo taken with you standing next to Ayrton Senna sometime.

On day two in the luxury hotel it was brought to my notice that whenever I stopped for refuelling in the bar, the Boss was loitering nearby looking ill at ease. Take notes, produce copy, interview someone, suggested Andy's pals. Which is why I have written 'Why is he watching me?' in my reporter's notebook. I got my own back by watching a film on the pay channel at his expense. I imagine it was supposed to be erotic but you never saw any of the working parts and from the looks of anguish on the young woman's face she appeared as likely to be being wormed as to be accommodating the empurpled manhood of her singularly unattractive escort.

So we watched the Supersport 400 Ultra Lightweight 125s race from Mr Bradley's garden before making our way to Signpost Corner for Sidecar Race B and the Classic Parade. If you wish, I will send you under separate cover a list of about fifty of those Classics that I would like to own.

Our Monday evening passed in a fruitless search for Joey Dunlop. 'Joey usually drinks in here,' we were told at every port of call. 'But he's not here tonight.'

In the morning I telephoned the boss and, catching him half asleep, talked him into allowing me the use of the company car. Common decency forbids me to tell you what cassettes he had left in the car. But I have a list.

In the late morning we set off in convoy for Ramsey and a day of Run What You Brung over the ⅛ mile. Arriving at the top of the descent into town, we found ourselves at the scene of an accident in which a young lad had hit a couple of cars at speed and died. Mac McDiarmid was with us and I was vaguely impressed by the hostility initially shown when he produced his cameras and before he explained that his photographs were not destined for the tabloids and another round of TT horror stories.

Retreating from this melancholy scene and reaching Ramsey by the back roads, we perched on the roof of a van parked overlooking the start line and owned by Kev. Thanks, Kev.

Now, I have a bit of an appetite for drag racing and can often be found looking pale and uninteresting at Santa Pod, drinking in extremes of noise and speed, but it is not quite like that at Ramsey. Mind you, there are numerous other delights for the casual spectator, amongst them a man who so dwarfed his machine that one feared that on the conclusion of an unsatisfactory run he might dismount and throw his bike out to sea. As a would-be shopper, I was much taken with the raffish lines and splendid noise of a '32 Rudge 500 but was told that these are in short supply just now. A pity.

We watched the Junior TT from Guthrie's, following a lunatic dash in the back of a pickup to get on station before the road was closed. As a country boy myself, I would have been happy enough just lying out in the mountain sunshine surrounded by amusing companions, fattening food and bottles of beer and wine. With the racing added I thought I should swoon clear away with pleasure. Sitting on the bank toying with a lager, I was reminded that this was dangerous work when Martin Ayles took his Yamaha rather wide and collected enough real estate for him to have required planning permission for a similar operation where I live in East Anglia. The dangers were yet more forcefully underlined when I looked up to see Steve Hazlett cartwheeling through the air, silhouetted rather beautifully against distant downtown Ramsey, to disappear out of sight behind a ridge. He must have been fifteen feet in the air when I first saw him and proceeding in a southerly direction at high speed. I'm glad you weren't too badly damaged, Steve – and I have a portion of one of the fence posts that almost performed unwelcome and high-speed rectal surgery on you if you want it.

Phillip McCallen and Bob Jackson, racing together behind Steven Hazlett, narrowly missed a generous portion of Yamaha still on the highway and marshals, conscientious as well as courageous, were still clearing debris twenty minutes later.

That night the rest of the party went to Onchan to watch Trevor Nation – known as 'John's mate Trevor' since I had my photograph taken with him – doing a bit of Stockcar racing. I elected instead to watch England play what they believed to be football against Denmark, the kick-off coinciding with an astounding display by the Red Arrows across Douglas Bay. I found myself in the absurd situation of shrieking 'Can't you bastards shut up and let me hear the match' as our brave boys shot in formation along the Promenade. I should have gone to Onchan. We all unwound later – we did a lot of unwinding that week – at the Pit Stop. 'There's Johnny Rea,' someone said, but before I could sidle over, notebook at the ready, for an in-depth interview, I was being introduced to Steve Hislop. 'Hizzy', as we insiders call him, looks about fifteen. How does he do that? Before I could discuss rose-jointed adjustable linkages with him I was being presented to a bored young woman who appeared to be drinking bubble bath liquid. 'He looks old,' she quipped playfully. I laughed hugely at this sally instead of giving her the glancing blow on the temple she so plainly deserved. 'Better old than ineffably stupid,' I chortled. She looked puzzled.

Outside, the streets were filled with the lightly pissed and the only Cockney voice I heard all week asked me for 50p. When I declined, it observed 'He's a c***!' to no one in particular, displaying the humour for which London is so justly celebrated.

On Thursday I moved to a different luxury hotel so that I could be closer to the Pit Stop. Before making the move, I had travelled with Andy Kershaw and some of the others to Jurby where Carl Fogarty was testing the Norton and where Andy hoped to interview Carl for *Down Your Way*. If you hear or have heard this programme, I want you to know that I almost died for it, a breakdown in communication having resulted in Andy being separated from his recording equipment by a good half mile and my having to run to retrieve the situation whilst Andy kept Carl happy gossiping about underbucket shim kits or something like that. Nature has not, in her wisdom, equipped me for running and I collapsed on the ground upon completion of my errand, the pounding in my chest indicating that my heart had decided that it had had enough of me and my foolishness and was planning to go into business on its own account having seen some likely premises in Rock Ferry. Only my sense of duty to *BIKE* kept me going.

Once Andy had been reunited with his recording stuff and I had scattered the paramedics, we set off around the northern section of the circuit in the pickup – with Carl Fogarty at the wheel. It dawned on me that my tribulations were not yet over when it became clear that Carl, who was being interviewed by Andy as he drove, was keen on demonstrating the racing line, oblivious to the fact that we were in a fourwheeler in rather less than racing trim and that there was serious traffic coming the other way. I couldn't scream or sob for mercy for fear of spoiling the recording and arrived back in Douglas with an even greater admiration for the men who fling themselves about the island for our amusement.

We watched some of the Senior TT from St Ninian's, where you can experience the thrill of having the bikes head straight for you at well over 150 miles an hour, and bluffed our way into the Grandstand for the finish.

Checking the films I just picked up from Boots, I seem to have taken eight photographs of Messrs Hislop, Fogarty and Dunlop on the winners' podium, but in a way I was more interested in Michael Noblett of Preston, who I think finished 18th, who, with all the flash and excitement going on behind him, drew up below me to show injuries he'd incurred in grazing a fencepost at Guthrie's to a small group of family and friends. Why does Michael, along with dozens of others, involve himself in this dangerous and costly activity with, I assume, little real prospect of success? I am afraid I was too embarrassed to ask him but as a spectator, I'm glad that he does. I hope we'll both be back on the Isle of Man next year.

As to the other question – what bike am I going to buy myself? – there is still no answer. The constant and grating repetition of a radio ad that started 'There's a current account that pays 9 per cent interest – but quite frankly, who gives a toss?' has hardened my heart against a Kawasaki. If I can't have the '32 Rudge, I want a machine without writing on it. Is there such a thing?

# Tubular Bells

*The Listener*, 7 June 1973

ON THE ALL too frequent occasions when I'm told that a record by a contemporary rock musician is a work of 'lasting interest' I tend to reach for my hat and head for the wide open spaces. Tony Palmer's theories notwithstanding, rock music, even the very best of it, is essentially ephemeral. Even when it does survive over any period of time – and rock hasn't yet amassed twenty years over which to survive – it is, happily, the stuff which the experts dismiss out of hand that makes it across the Styx. In the fifties the pundits predicted that the Presleys, the Cochrans, the Little Richards and the Fats Dominos were merely transitory phenomena and would fade away with the first onrush of reality. The really durable music, they cried to a man, was the jazz and serious pop of the period. Well, forgive me, but I don't even remember the names of the people they praised for their quality and durability.

Today these same experts or their descendants would probably tell you that in twenty years' time collectors will still be enthusing over the records of such weighty bands as Yes and Emerson, Lake, Palmer. I'm ready to bet you a few shillings that Yes and ELP will have vanished from the memory of all but the most stubborn and that the Gary Glitters and Sweets of no lasting value and *Top of the Pops* will be regarded as representing the true sound of the seventies.

Having said that, I'm going to tell you about a new recording of such strength, energy and real beauty that to me it represents the first break-through into history that any musician has made. Mike Oldfield even had the nerve to be something of a child prodigy, recording a pleasant, if somewhat sugary, LP of dainty songs with his sister Sally at the age of fourteen. A year later he was working alongside such seasoned professionals as the itinerant sax hero Lol Coxhill and the composer David Bedford in Kevin Ayers' erratic but witty and articulate group, the Whole World.

In 1971 Mike Oldfield began work on a composition entitled *Tubular Bells* and now, after 2,300 overdubs, *Tubular Bells* is available on record as the first release from Virgin Records of Notting Hill Gate. Too often have we read of music that 'crossed the barriers between pop and the classics' when it manifestly did nothing of the sort. For too long also has the business of reviewing records been a routine of dusting off superlatives and arranging them in attractive sequences. With *Tubular Bells* we have a record that does quite genuinely cover new and uncharted territory. Without borrowing anything from established classics or descending to the discords, squeals and burps of the determinedly avant-garde, Mike Oldfield has produced music which combines logic with surprise, sunshine with rain. In the process of so doing he plays a bewildering range of musical instruments without ever playing merely for effect. Each device is there because that is where it should be. On the recording he has some assistance from other musicians, and it must be an indication of the scope of *Tubular Bells* that the forthcoming live performance on 25 June at the Queen Elizabeth Hall in London will draw on the talents of Steve Winwood, the curious Viv Stanshall, Robert Wyatt (my own favourite drummer), the aforementioned David Bedford and Kevin Ayers, Frank Ricotti and members of Gong and the underrated Henry Cow. Sadly, I can't anticipate much interest from

Radio 1, but perhaps it is not too late for Radio 3 at least to arrange for this first live performance of *Tubular Bells* to be recorded.

# Turntable Mistress

*Observer*, 22 December 1985

---

WE HAD been welcomed by Paula Yates and Radio 1 disc-jockey Simon Bates to the Hippodrome, Leicester Square, and the National Final of *The Star*/Babycham Female DJ of the Year Competition 1985. Simon is part of the prize for the winner, who gets to spend ten days with him in the United States before enrolling at University College, Cardiff, for a three-month radio training course.

Before the contenders started to, er . . . strut their funky stuff for the folk on the dance floor, Simon and Paula introduced them to us. One was an ex-Playboy Bunny, another worked part-time for Eastern Electricity, one wanted to go to New Orleans, another to own a bistro.

'This is Madonna. Wooooo! I want to see you dancing,' squealed Stephanie Callister (23). Stephanie was the first of the runners out of the stalls and didn't wear very much clothing. She was heavily photographed. At one point, and doubtless overcome by the importance of the occasion, she yelped, 'Camera, make love to me.'

Jakki Gee won the North-West of England Regional Final at Manchester's Millionaires Club. Jakki wore a lot of clothing, albeit clothing which tended to part to reveal something of the essential Jakki beneath. She was unique amongst the contestants in introducing an element of carefree humour into her turn. The audience looked puzzled.

Chrissie Jackson of Norwich, demurely dressed, started well with a bit of a 'rap' in the modern manner. After this she fell back on a device used by many of the competitors – standing about, looking embarrassed.

Pre-match gossip had it that B. J. Stevens, who worked as a DJ in Germany with Bacchus for five years, was a likely winner. She has an enviable wealth of blonde hair which she now tossed, now brushed back from her eyes. Sadly, once she had done this a few times, she joined the ranks of those who could only stand and stare.

Fay Williams, the final contender, won my support by playing a Fall record. This virtually cleared the dance floor. Fay was born and educated in Somerset, gaining eight 'O' levels and two 'A' levels.

The commonly held view was that as Trish Roberts had been the only contender to prod the paying customers into anything approaching life, urging them to involve themselves without delay in a 'boy girl, boy girl situation' which turned out to be a conga, she must be the winner. I hope she enjoys Bates, the States and Cardiff.

# UK Fresh 86

*Observer*, 27 July 1986

---

'IF YOU'RE chillin' in the house, say ho.' Ho, we went but without conviction. Much of the afternoon had been given over to saying ho and the modest appeal of this severely unfulfilling activity had all but evaporated.

UK Fresh 86, an explosion – let's settle for that – of hip hop electro music, was the first event of its type in Britain. Hip hop has increased in popularity here over the past year to the point where it is regularly heard in the charts and in television commercials, so this was the right time for Morgan Khan, high profile head of Street Sounds Records, whose twelve hip hop compilations have kept British enthusiasts abreast of the music, to promote a celebration.

Wembley Arena may not, due to its size, have been what the B Boys, B Girls and this B Middle Aged Man would call the place to be. For the afternoon show a third of the seats were not taken and although the avid consumers flocked to the stage, there was still an impression of emptiness.

UK Fresh 86 started well enough, with the duo Word of Mouth and their man on the wheels of steel, DJ Cheese. Word of Mouth were no sucker MCs (are you with me still?) but Cheese was awe-inspiring, scratching and cutting records at dizzying speed whilst cueing others, leaving his turntables only to perform briefly as a human beatbox.

Soul purists decry hip hop, arguably the only original music of the 1980s, for its repetition and reliance on mechanical aid. I rather suspect that in thirty or forty years' time collectors will discuss hip hop in the awed whispers they now save for the blues of Mississippi and Chicago. The outcry of an oppressed people? It certainly can be. Although far too many of the records are wearisomely sexist, some of the best have been sharply political.

Although Dr Jeckyll (sic) and Mr Hyde were amongst those who loved UK, they kept the ho-ing to a minimum and their performance sharp, funny and uncluttered. Three headline acts were, in truth, rather dull, wearing costumes that would have brought a whistle of disbelief from Elton John, and behaving both on and off stage in a manner more reminiscent of early 1970s rock stars than of Mean Street survivors. Grandmaster Flash even urged us to shout out our zodiac signs.

# USSR

*Observer*, 6 March 1988

---

MOSCOW'S Olympinsky Village, internally clad with light coloured woods and boasting all the invitations of rock 'n' roll untidiness that one associates with the Purcell Room, was playing host – for one night only – to the local rock community, on this evidence a parcel of substantially less than mutinous-looking young people. Some of the youngest were accompanied by fathers radiating liberalism and understanding. Low-key security was provided by recently shaved lads in dark blue suits, collars heavy with dandruff, who looked on with reasonable concern as the first troupe of entertainers, its members wearing foolish costumes – the lead singer was dressed as for *The Desert Song*, the (male) guitarist as a woman, a percussionist had a cardboard leek gummed to the top of his head – appeared to slight applause.

Clearly deriving from a theatrical tradition rather than from the rockaboogie equivalent, the group reminded me of those squads of hippies who, in the late 1960s, had me wondering whether the time had finally come to cast off my shackles and renounce peace. When I spotted that another guitarist had a large key issuing from his back, I quietly slipped my interest into neutral. If there had been a bar I would have sought it out.

Next on the bill of fare was a duo who, Russian friends told us, utilise pre-revolutionary cabaret songs and styles to satirise the present regime. They do this with check waistcoats and false noses, the singer carrying on in a manner that marked him as the Morrissey of pre-revolutionary cabaret. Sadly, I missed much of their set as I had been led away by a pair of the blue-suited security boys eager to learn why I was videoing the proceedings.

As I apologised again and again for my inability to understand the Russian they were shouting at me, so more shouters were summoned. Ultimately I was at the centre of a circle of eight or nine of them and we provided, I felt, a better turn than we had so far seen on stage. (Anyone seeing this incident as the mark of a harsh, intolerant society should try videoing a concert in Britain and see what happens.)

Zvuki Mu topped the night's bill. They enjoyed a considerable local reputation and their appearance on stage finally silenced the most dedicated of the shouters, a zealot whose only word of English, which he repeated often and at impressive volume, was 'administration'.

Zvuki Mu are not, by Western standards, terrifically rocky in appearance. Soberly dressed, as though for job interviews, they look like rather decent ornithologists, albeit with a lead singer, Peter Mamonov, whose behaviour was reminiscent of that of the great Roger Chapman, once of Family. The songs, impenetrable to us, were essentially morose, although punctuated with sudden and rather wonderful squalls of instrumental violence. Their best piece, to Western ears, was a Velvet Underground-esque work of considerable intensity. The audience was sufficiently impressed with Zvuki Mu lyrics to applaud individual lines from songs.

At the end of the concert the chief amongst the shouters reappeared, but only to apologise for having disturbed my evening.

# Virginity

## Virgin on the Ridiculous

*Sounds*, 4 June 1977

---

YOUR FAVE weekly. Last week. Made a bit of a resolution, didn't I? 'I promise I won't mention football again until next season.'

You must be joking.

Did you see them? I mean, DID YOU SEE THEM? Like gods they were, every one a super-hero. Even without Keegan, do you think anyone is going to be able to stop them next season? And the Liverpool supporters – dey wus magick tu. I didn't hear a single syllable of German all night.

After the game, when we were all sobbing joyously into the champagne substitute, the phone rang and rang again as all manner of folk, even Ipswich supporters from down the lane, phoned to congratulate me. You'd think it had been me and not Tommy Smith who'd soared like a brick gazelle over the defence and howitzered the pellet into the net for the decisive goal.

If there are going to be Jubilee honours, then I look for Sir Thomas Smith, Sir Robert Paisley and Sir Emlyn Hughes – and that's just for starters. Give the usurers and gangsters a miss this time, Your Majesty, and get down to true nobility.

Speaking of the Queen. Regardless of whether you consider

yourself a punk, a hippie, a rocker or whatever – and, whichever way you look at it, you're probably just part of some vicious bugger's global marketing exercise – you're going to have to face the fact – later, if not now – that the Sex Pistols single is one of the great rock records of all time.

I was just about to suggest that it ranked alongside such ancient faves as 'Summertime Blues', 'Long Tall Sally', 'Satisfaction' and 'My Generation'; then I thought – not easy, 'cos I've still got the dying echoes of a hangover rumbling around the frontal lobes – 'What, Peely, old scout, is the point?'

Why is it that we elderly chaps like to take solace from murmuring, smiling indulgently the while, that the 'new-wave' is really nothing new at all? 'Sound just like The Who at Eel Pie Island, don't they?' And that despite the fact we never saw The Who at Eel Pie Island.

Perhaps it all helps us convince ourselves that although we're too pooped to pogo, we really do understand what's going on. And, *ma foi*, we sympathise too. Certainly.

I probably don't understand at all, but

(1) I don't give even a small portion of a fuck, and
(2) 'God Save The Queen' is a devastating record – and this time I do mean 'devastating', and
(3) this year I have added more good records to my collection than in any year since 1955.

To change the scene slightly I'm writing this in the Radio 1 typing pool again, only a few feet from where a Certain Producer said last week that he would not want his daughter to buy a record with the word 'punk' in the title. Great stuff, eh?

And someone from the Fourth Floor – where the real power is concentrated – has this second come in and asked me to try and recall my Most Embarrassing/Humiliating Moments for a special Noel Edmonds (remember him?) Jubilee Prog.

Here's one I'll not be able to use on the radio. It is a whit embarrassing to admit it to sophisticated young persons* such as yourselves, but I retained my virgin status until I was well past my twenty-first birthday. You should have seen my wrists though – like pillars of mighty oak. I was unbeaten at arm-wrestling for seven years.

Well, I finally surrendered my body to an exceedingly grimy woman of some thirty summers.

The episode, as you might imagine, was clumsy, deeply humbling and, I fear, less than fulfilling for the luckless woman, who was understandably reluctant to believe that a warrior of my seniority was quite unversed in the arts of lerv.

In the months that followed this less than rhapsodic defloration, I somehow managed to persuade myself that I was a bit of a catch so when the chance came for me to carve a second notch on the handle of my gun (*ugh! – Ed.*) I approached the task with the wonderful confidence that is born of ignorance.

Soft music played (I think it was a Dave Brubeck LP, so that

* For 'sophisticated young persons' read 'dirty little bastards' throughout.

should give you an idea of what sort of twerp I was), the victim had been wined and dined (cheeseburgers and Dr Pepper '61), and the necessary medical equipment had been obtained from the Texaco station on the corner.

For several minutes I worked away, eyes closed with the sheerest rapture, certain that the young person beneath me was transported with delight at the skill and fervour of my love-making, before it occurred to me that she was keeping her passion pretty quiet and pretty much to herself.

For a moment I opened my eyes, allowing myself a glance at the sofa on which we were sporting and at the fervent creature who was reaping the manifold benefits of my body. She was reading a magazine!

Must pop off now. Before I go, can any of you identify the young woman in the picture? And what is she doing sending me photographs of herself dressed in this lewd fashion? I don't want to do anything to spoil your enjoyment of Jubilee year, but her clear resemblance to our lovely Princess Anne leads me to suspect that she may herself be of the House of Windsor.

# Voice-overs

*Radio Times*, 7–13 January 1995

---

RECENT REVELATIONS about the monies paid to the multi-talented Chris Tarrant may have led you to believe that we have similar sums being delivered to our front door by a chain of men with wheelbarrows. 'Of course,' your reasoning may have run, 'he's only half the DJ that Tarrant is, so he probably gets about half as much money.' Pause for hollow laughter. Not that I am complaining about the wages I get for my radio work for the BBC and others. It is more than enough for our simple needs and ensures that we can drink wine with our meals and can replace the children's trainers on demand.

However, I do do voice-overs for television commercials, the media equivalent of taking in washing, and was therefore interested to see a preview of Thursday's *Situation Vacant: The Advertising Executive* on BBC2. The first voice-over I did was about fifteen years ago, when I was, briefly, the voice of the Marmite Baby before being supplanted by Willie Rushton. Since then I have murmured apologetically on behalf of stout, fizzy drinks, chocolate bars, toilet paper, tinned fruit and a lawn treatment which, if a shoal of incoming hostile mail was to be believed, contained an ingredient banned under the terms of the Geneva Convention.

I have turned down as many as I have accepted though and

have, as have all those who similarly toil in the recording studios of old Soho, done several perfectly fine voice-overs that were never used. We had a particularly gruelling Christmas about five years ago when, in every commercial break, up would come one of the three ads I had done only to be replaced before transmission by that Griff Rhys-Jones. I saw this Jones at a BBC Christmas do only a couple of weeks ago and was, even now, tempted to kick him firmly at the base of the spine. Only cowardice kept me from manhandling the wretched fellow.

Now, doing these voice-overs may seem to you something of a doddle when compared with, say, air-sea rescue or steel-erecting but, let me tell you, real suffering is involved nevertheless.

When you arrive, all of a fluster, at the studio, you are shoved, after a delay timed to remind you of your place in the scheme of things, into the presence of the various people responsible for the commercial. You are, for an hour, their hireling and they know it. There are never less than six of them, often there are more. They are invariably beautifully, if casually, dressed, bright, amusing, attractive and haven't seen each other since ... was it Grenoble? I am none of these things and have never been to Grenoble.

After a round of introductions so intimidating that you remember not a single name, although you think one of them might be called Piers, the advertising dreamboats prod you into the ill-lit booth in which you are to work. As the door closes you can hear them laughing. After your first tentative stab at reading the daft words they have been crafting for the best part of a week, you can see them arguing through the glass that separates you from them. They are saying, 'Whose idea was it to hire this twerp?'

Alone and near to tears in the booth, you realise that you're not even sure which country Grenoble is in.

Eventually the advertising executive delegated to speak to the staff will press the talkback button and say, 'That was fantastic, John, but . . .' In the language of advertising, fantastic is very, very bad indeed. After an hour of this, you emerge giddy with self-loathing but knowing that little William or Danda or Thomas or Florence can soon have some new jeans.

In *Situation Vacant* you can see how these people are selected for employment by the agencies. Sean, Maxine, Giles and Hazel have to work together on a campaign for the Spastics Society while being observed by the BBC2 cameras and the Head of Human Resources at the agency they wish to join. It goes without saying that Sean, Maxine, Giles and Hazel are beautifully, if casually, dressed, bright, amusing and attractive. They are also panic-stricken, knowing that, even as they collaborate, they compete. I watched the first half, then, when I returned a day later for the dénouement, found the videotape missing. It is probably even now on the sitting-room floor with an episode of *Home and Away* recorded on it, so I shall have to watch along with the rest of you to discover what became of Sean, Maxine, Giles and Hazel.

# Loudon Wainwright

*Disc*, 1970–1

‘I LOVE HIM, I really do,’ said the girl who hadn’t stopped talking since he started playing.

‘Some of his lines are all right but there are too many verses,’ said the groover standing behind John Walters, the Pig and me.

‘Who is this guy?’ asked someone over at the bar.

‘Loudon something – never heard of him.’

‘This is crap,’ said the groover’s friend firmly.

You’ve probably seen one of those films about hardened criminals escaping from a Southern prison, eluding their clean-cut pursuers for 70 minutes or so before being cornered – usually in a swamp – and being returned to captivity so that we can all sleep easily again. In the course of the action you’ll probably recall that one of the fugitives gets shot and that he tends to be the youngest, the one with the twitch and the crazy, pale eyes. On stage Loudon Wainwright looks pretty much like him.

In a situation when superlatives were debased long ago and superb means ‘well, OK’ and ‘major talent’ means ‘O’ level English, you’re not going to believe me much when I tell you that Loudon Wainwright is a superb major talent. His first LP has been in the flat for several months now and it has slowly dawned on us that here is something rather special, that rare notch above the excellent.

On stage he looks awkward and very much out of place. He's the least groovy-looking guy you'll have seen singing for a long time; his long tongue rips out and licks his lips feverishly and frequently, he bares his teeth like Kirk Douglas, his foot stomps and sometimes he smiles a smile that seems to lie somewhere between mirthless and madness.

I have never seen a more powerful artist. His only equal in my experience has been Son House and the groovers talked through his set too – so much so that, for the first time since I was seventeen, I was moved to offer violence.

When Loudon Wainwright's is a fashionable name, and you're paying huge sums to see him at the Albert Hall, the people who giggled and chattered through his set will be reminiscing of the night they saw him for sixty new pence (fifty for members), and they'll be flocking to the current trendy house of horrors to buy atrociously recorded bootleg LPs of Loudon Wainwright concerts.

His songs are based on New York City in particular, New England in general, but you'll have no trouble identifying with them if you'll only listen. His voice is high, broken and lonesome. Last week we exchanged hellos in the Atlantic offices and I know that, despite myself, I'm going to be bragging about that one of these days and telling fictitious stories about 'my friend Loudon'.

I won't try to tell you more about him but, if you know and trust me, just try to hear him. Loudon Wainwright III is someone very rare and remarkable.

'He can't sing but the guitar playing is pretty good,' said the groover's friend, and John, the Pig and I fled into the night. When I was about eight my mother introduced me to the work of Paul Klee and I've loved his paintings and drawings ever since. Driving away from Hampstead I felt rather as I would if I'd seen someone writing the address of a boutique over 'The Twittering Machine'.

It's not that I don't like doing DJ gigs, it's just that something always goes wrong. Before going to the Poly they'd assured me that there'd be two turntables with lots of power. 'Don't worry a bit,' they said. Sometimes when you get there they say, 'Jesus, I knew we'd forgotten something' and shout 'Hey, Phil, do you still have that old record player?'

'Wow, man, don't worry, Phil's great at fixing things up and we'll find you a deck somewhere and Phil has this amplifier he made out of refrigerator parts and we'll use the group's p.a. and everyone's here to get pissed, anyway, so it doesn't matter much.'

Well, it wasn't that bad but it didn't work and, of course, it does seem to matter because not everyone is there for the beer and I spend a lot of time looking for records to amuse and divert. What I must do is find a strong and powerful round of disco gear ordinaire, surround myself with congenial companions, and set forth to do it all properly for a change. Jugglers, fire-eaters, dancing ladies. If you're one or more of these you might let me know.

Most of the people were nice, they always are, but there's always one who gets very drunk and tells you: 'Hey, just 'cos you're on the radio doesn't make you better than me, you know.' You say that yes, you do know that and he leans on you a bit more and says it again only louder and then asks you for your autograph 'for my girl, she's too shy to ask for herself'. So you laugh a bit and say well, autographs are a bit silly really, aren't they, and he starts to get angry.

Another archetype just has to let you know that he's stoned, in fact so is his chick, in fact they've been stoned since Wednesday, in fact is it cool if they roll one now, in fact it's only Turkish, in fact, in fact, in fact. Wow! Where, you wonder, is someone who is being real and when will means no longer be ends?

Thankfully it's not like that at the Nag's Head. People don't talk as much as they used to but they're probably just used to seeing me lurking in dark corners and never imagine that people, especially people who know Marc Bolan and other famous people, need talking with too. Still, those Friday evenings are things I look forward to all week – it's comforting to see Bob, Pauline, Anthony and Paul and know that they're quite pleased to see you even if Anthony is going to thrash you on the football machine.

Sometimes it's Portsmouth, the Tricorn Centre and Cromat and that's something else to look forward to as well. It's a funny sort of place, the people are friendly enough though and sometimes we play football on the parking lot before the club opens, even though Ricky isn't as fleet of foot as he might have been once upon a time. The ladies, as I've said before, are pretty in Portsmouth and have the kind of distance to them that goes with knowing it. They don't need to have, of course, but they haven't learned that one yet.

I must get that disco gear and learn a bit about punctuation too.

# John Walters

*Radio Times*, 11–17 August 2001

---

HEAVEN ONLY knows there are enough deities for us to choose from and those we worship already seem to cause an awful lot of trouble, so this may not be the best time to introduce you to a new one. He – I'm afraid it's a he again – is Snibri.

As is so often the case with these things, the origins of Snibri are a little obscure but have to do with the spectacularly untidy office presided over by John Walters in the former Radio 1 offices adjacent to Broadcasting House. There was something inspirational about this office and the mountains of records, minutes of pointless meetings, demonstration cassettes, promotional material, BBC handbooks, unread magazines and portions of machinery long since divorced from their true context but kept against the day when we might rediscover their purpose. From time to time some localised tremor might dislodge something from these mountains but, by and large, they maintained a remarkable stability.

They also infuriated the then controller of Radio 1, a man with an uncommonly orderly mind, who alternately begged Walters to tidy his office or threatened him with redecoration that would sweep all the debris away for ever. Walters smiled cooperatively and did nothing.

But where does Snibri come into all of this? Well, Snibri is a deity devoted, above all else, to chance and unfair advantage. It was, after all, Snibri who ensured that my name remained for years on the CBS (now Sony) mailing list, long after we had stopped playing any CBS records. Every month a vast shipment of essentially unlistenable records would arrive in the office and we would breathe a silent prayer to Snibri. Occasionally Walters would decide it was time to render unto Snibri, as it were, that which was Snibri's, and some object would be given special recognition within the chaos. An example was the elaborately wrapped parcel which Walters stopped trying to penetrate after four or five minutes' vigorous work with a pair of blunt scissors (a Walters office would come equipped with no other form of scissor). This parcel was deemed worthy of Snibri and remained unopened in the office for at least a decade.

Just a week after a previous column about John had reached homes and newsagents around Britain came the news that he had died unexpectedly in his sleep. We had worked together for Radio 1 for more than twenty years until ill health forced his early retirement. I owe Walters more than I owe any other person in my life. He taught me that there was nothing shameful in getting things wrong from time to time, provided you remained true to some sort of ill-defined but genuinely held principles – and popped around the corner for a beer if time permitted.

Whenever I have received an honorary degree or similar tribute, I have known that no more than a third of it was really mine, with a third going to Walters and a third to Sheila, my wife. Today I feel as infantrymen in the trenches must have felt when the man beside them was hit.

# Wham!

*Observer*, 6 July 1986

'WE'VE GOT FOUR years of thank yous to say this evening,' said George Michael. We whooped appreciatively. In the five hours since the Wembley Stadium gates had opened we had done the World Cup wave not once, not twice, but a dozen times, with all the giddy abandon of holidaymakers. We had been entertained by Gary Glitter, wrapped in silver like an oven-ready turkey, in a set which had had Gary striking a series of heroic poses and bellowing his sounds of the seventies.

The audience, older, drunker and ultimately pinker – 'Get yourself a great tan,' the compère had urged – than I had imagined, was in high spirits. Beautifully presented, ready, one felt, to travel and meet people, they clapped along fervently.

'Is this the sort of music you like, Daddy?' Alexandra (8) had asked, and I had assured her that it was not.

The immense stage, great grey sails trimmed with black and cut to give the impression of a ceremonial helmet of oriental design, next played host to Nick Hayward, performing live for the first time in two years. The problem with playing music in stadiums is that it can end up sounding like stadium music, booming and empty. The audience applauded their memories rather than the present performance.

As curtains bearing the legend 'The Final' closed across the stage, we were treated to an exclusive preview on the twin video screens of the film of Wham!'s trip to China. This was pretty dull fare, a gaudy home movie which proved little beyond the awfulness of being followed everywhere by cameramen.

On the covered pitch fans unfurled a banner which read 'Knob out George'. They were addressing a leaner, meaner George Michael. Gone was the Princess of Wales haircut and the suggestion of a corporation; this was a *West Side Story* George Michael, all black leather and flashing eyes. He ran out on to the catwalks which extended some distance into the audience, pausing to wiggle his bum and for us to scream our screams, setting the pattern for an evening which involved a lot of shameless cheerleading and much athleticism, before retiring to the sidelines to allow Andrew Ridgley to do much the same thing.

These formalities out of the way, George grabbed a microphone and Andrew grabbed a cosmetic guitar, effectively bowing out of the proceedings apart from some brief moments tinkering with tubular bells late in the night.

George Michael has a strong pop-soul voice and he writes vexingly catchy songs, the best known of which were paraded before us over the next two-and-a-half hours. Wham! have survived where others have not simply because they have kept working, never making the fatal assumption that they could lay off for a year, then pick up where they left off. At Wembley the paying customers revelled in the duo's success and in George Michael's ability to play upon them as upon a stringed instrument.

They never wearied of the flirtations and the running about, and when Elton John appeared in an asinine clown's outfit – does he do these things to mask some insecurity? – they quite correctly perceived that this was an honour for him rather than for Wham!

We sang along, we waved, we stomped for encores we knew they would do anyway. There must have been times when George Michael thought it was all far too easy. At one stage he started, unexpectedly, to speak about the Press. 'You've read the reports about Mr Ridgley and myself . . .' he said, but the moment passed and he went back to cheerleading.

Later George Michael told us that this was the best day of his life, but the clear implication was that there are, for him at least, better to come.

# Whatever You Want

*Radio Times*, 10–16 April 1999

---

JO, ZOË and Joanna love whales. They're huge, gentle giants, rhapsodises Jo. Mesmerising species, suggests Joanna, showing off a bit. Zoë claims she'd rather like to be a whale. In order to win the opportunity to join other whale fiends on a boat in the Norwegian fjords, Jo, Zoë and Joanna have to sit in a sort of whale-like vehicle and bounce up and down. By doing this they activate a hidden mechanism that reels in a baby whale (simulated) on the end of a rope. Whoever gets the baby whale reeled in first wins the trip to the fjords. Jo, Zoë and Joanna are contestants on *Whatever You Want*, presented by all-smiling, all-shouting Gaby Roslin. *Whatever You Want* is breathless stuff and no place for the unassuming or the modestly reflective. In what Gaby describes as a completely bonkers new game, competing couples – couples in that there are two of them, rather than couples who are united in holy wedlock – ride on that terrifying roller coaster in Blackpool while carrying open pots of yogurt. Those couples who retain the greatest volume of the popular dairy product in their pots go forward to the next round, next week. The eventual winners will get a round-the-world trip complete with spending money.

If you find yourself short of ideas on how to explain life in Britain in 1999 to, say, a visitor from Uzbekistan, *Whatever You*

*Want*, now into its third series, it seems, would be a fair place to start. I mean, you must have noticed, for example, how no event staged anywhere in the country can be considered worthwhile unless a celebrity is present. Your celebrity may be an Anglia TV weather forecaster or it may be yummy *Top of the Pops* presenter Jamie Theakston, but your function must be validated by the presence of someone vaguely identifiable. In the case of *Whatever You Want*, it's Jamie Theakston and he is there, locked in a stretch limo, to be won by whichever of three pubescent girls can guess, among other things, Jamie's inside leg measurement. All three contestants, when asked what it is they most admire about the swoonsome entertainer, cite his hair. It may just be that having little or no hair myself has unseated my reason, but regular televiewing could convince you that there is nothing more important in life than hair. You will have seen, I expect, that advertisement that gives you advice on enjoying your hair. I don't remember that when I had hair myself I was ever conscious of actively enjoying it. Coming out for a beer, John? No, not tonight. I think I'll stay in and enjoy my hair. (To those of my critics who have derided the woolly hats I have been seen wearing in *Sounds of the Suburbs*, currently beating a retreat through the Channel 4 Saturday-night, Sunday-morning schedules to a chorus of 'and in next week's programme, at the slightly later time of . . .' let me explain that I wear woolly hats to keep my hairless head warm, not to be cute. Trust me.)

'Give our love to Zoë,' bellows Gaby as Jamie, much more than a nice haircut, if a couple of hours in his company in the bar of a Birmingham hotel last year are anything to go by, leaves with the sobbing winner. Gaby is invoking the name of Jamie's co-presenter of *Live and Kicking*, Zoë Ball, 1999's ultimate celebrity. I must find out where Zoë is going to be on New Millennium's Eve. Such a conjunction of time, place and celebrity must surely trigger something unforgettable.

Another bonkers competition offers another sobbing child the opportunity to travel to Belize – 'a creepy-crawly paradise,' roars Gaby – to help scientists collect what are described, confusingly, as some of the most varied species of insect. Yet another brings us three Tyneside women who want to blow up a tower block. Give me the chance to do this, with the celebrities of my choice locked inside the building, and I do believe I'd have a go on *Whatever You Want* myself. The three women are blindfolded and hurl handbags at cardboard tower blocks, of course. Everyone whoops continuously and, if given the opportunity to do so, jumps up and down with excitement. Three young men play *Give Us a Twirl* for a chance to go white-water-rafting in the Grand Canyon. I suppose we should console ourselves by considering that if *Whatever You Want* was Japanese, the whale fanciers, the rafters, the Theakston freaks and yogurt carriers would have mice in their underclothes. Maybe in the next series, hey?

Taking some moments off from enjoying my hair to play a bonkers new game I've called *Give Us a Break*, I sat down to enjoy programme one of a new series of *The Adam and Joe Show*. Having enthused over the two previous series, I'm reluctant to say much about the new one for fear that cynics among you – and I know you're there, you rogues – might believe that I have a financial interest in the programme, but the toys version of *Saving Private Ryan* is good and there will be those who are entranced to hear that the first pop personality under investigation by the Vinyl Justice Squad this time around is Mark E. Smith of The Fall, a man who makes Adam and Joe look conventional. What a hero. Where's Mark going to be when the millennium cracks? That's where I want to be.

# WI

*Radio Times*, 27 October – 2 November 2001

I JUST hope they'll be gentle with me, that's all. After all, I've never done this sort of thing before. Oh yes, I've been tempted. Of course I have. I mean, it's not the first time anyone has, you know, asked me to.

But, hey, enough of this heavy-handed coyness. The thing is, I have been invited to talk to a local Women's Institute and I haven't a clue what to say to them. Talk for thirty-five minutes, someone suggested, allow another ten for questions and answers, then make your excuses and leave. Easy to say that sort of thing – and my friend John Walters, who could talk for at least that length of time on the cue 'Hello John,' would not understand my anxiety at all. The women will not want me, reason insists, to talk about, say, the recorded work of Japanese noise typhoon demons Melt Banana. Or the merits of drum 'n' bass overlords Stakka and Skynet. So what do I talk about?

Cooking is very popular but I don't cook, unless you count gas-house eggs. I do eat, though. Always have done. Rather too much, in fact. Perhaps they'd enjoy my account of the problems encountered by a vegetarian in the Europe of the late sixties. In those days, cooks regarded vegetarianism as, at best, deviant, at worst, an insult and would get back at you by putting diced ham in almost everything. 'But it's only ham,' they'd say, if you demurred.

The last time Sheila and I ate meat was about ten years ago in Moscow. We had not eaten meat for ten or fifteen years before that but these people had invited us for dinner and served us with what looked like overcooked fists. God knows from which of his creatures the meat had been hacked, but we knew our hosts must have gone to a lot of trouble and no little expense, so we ate with, I hope, every appearance of gusto. How we'd have got by in Roman times, when we'd have been served a whole roast pig stuffed with nightingales, I can't imagine. But that won't take up thirty-five minutes, will it?

If I knew more famous people, I could tell them highly coloured stories of the stars, but I don't imagine they'd be bowled over that I said 'Hello' to Mark Lamarr last week. I was pretty pleased myself, and several people to whom I have mentioned it were impressed, too, but Norton Women's Institute? I don't think so. Amusing on-air moments? Well, last night's programme was a bit of a collector's item. The thirty-eight-minute recording of the White Stripes in concert stopped for no discernible reason after seven minutes and I had to restart the CD in a different machine. Then I played a trailer for Radio 1's upcoming week in Birmingham instead of the news-jingle, the newsreader played the wrong insert for the lead news item and I had left a record playing, with the result that grim stories about death in Afghanistan were punctuated with sardonic laughter. No, that's not good enough, either.

I've got another four days to think of something. If anyone from the Norton WI is reading this, I'm sorry I was so boring.

# Wine-tasting with Walters

*Sounds*, 24 June 1978

---

LISTEN, YOU rabid self-defilers! You may well consider that much day-time radio is pretty rank, but just be grateful that you don't have to watch day-time television, that's all I can say. Unless, of course, you do.

I've just switched on the machine ('the electric governess', as Dirk Hamilton has called it on his pretty good Elektra album *Meet Me at the Crux* – available on import) and damn me if there isn't some chap on there trying to teach me how to say 'goat'. I like to think that, having endured ten years of expensive private education in some of the most violent schools in Britain, I can say 'goat' as well as the next person. 'Goat' – there, I said it again. Kinda cute title for a song, hey? But perhaps this goat-laden programme is not intended for me.

One supposes that the broadcasting authorities have to find something with which to fill in the yawning time-chasms between football matches, but why cannot they bring us film of, say, Siouxsie and the Banshees or the UK Subs rather than educational featurettes for friends of the goat? Last week I surprised young Winner as he watched *Play School*, and I sat him right down and gave him a good talking to, you bet. You may think I'm being overly severe on the boy, but no one who is going to be playing for Liverpool in another fifteen years can afford to fritter away their time learning how to construct petrol stations from discarded egg-bixes. All right then, egg-boxes. Pedant!

I'm very much afeared that my choleric mood this afternoon may not be unrelated to the considerable quantities of wine I drank the other day. My system, admired though it is by all those who are familiar with it, hasn't yet fully recovered from the cruel onslaught of the gallons of fine Hocks and Mosels that were forced on it last Tuesday afternoon. Unwisely perhaps, I had had no luncheon and only a chota-hazri upon rising, and the effects of the wine were immediate and considerable. I had been invited to a wine-tasting by John Walters, who will doubtless write of the same event in the next issue of *ZigZag*.

(He may well also advise you in this same magazine that I shave my armpits. Quite how he conjured up this particular diseased imagining I don't know, but he has been spreading the fiction about with considerable enthusiasm and his own unique inattention to detail. I'll bet he won't be so quick in telling *ZigZag* readers (if there are any) that so humiliated is he at being a sufferer from Tormenting Rectal Itch, as advertised in the best periodicals, that he sends me out to buy his Preparation H. There!)

Now where was I?

I suspect that letting slip the information that I have been to a proper wine-tasting may destroy what remains of my street credibility, and I have to face the possibility that Jimmy Pursey will never speak to me again, but I'm bound to say that I found the whole event rather wonderful.

My own favourites were a 1976 Wintricher Ohligsberg Riesling Auslese and a 1977 Bernkastler Badstube Kabinett, and by the time I had sampled plenty of these and compared them, for scientific purposes only naturally, with a dozen or so other wines, I found myself saying, in all seriousness, that I found another wine – I think it was the '76 Zeltinger Himmelreich

Riesling Auslese – rather 'leafy'. I do appreciate that 'leafy' is not a part of the vinophile's largely inaccessible vocabulary, but I knew what I meant when I said it. When I tell you that Dick Jewell, the *Sunday Times*' man on the fringe of rock, is also their wine expert, then you will know in what dangerous waters I am cruising.

What with one thing and another, we lingered perhaps a moment or two too long at the wine-tasting, and several hours later the noble Peel head was still aswim with rich fumes when its owner/operator was refused admission to Dingwalls, Chalk Farm's fashionable dance venue. I was there to introduce a performance by George Thorogood and his Destroyers. Once I had gained entrance I'm sorry to say I behaved poorly, consuming yet more wine and making an incoherent speech from the platform, a speech which included my by-now-traditional attacks on *The Old Grey Whistle Test* and Capital Radio.

I can't tell you exactly what I said, but whatever it was it grievously offended some Capital Radio fancier who was present – so much so that I suspected for a while that he was going to biff me one. Happily he didn't. Neither did Roger Scott, the Capital Radio afternoon personality, who turns out to be, oh, about *that* tall and from whom a biffing would perhaps have been a melancholy business for the biffee. Happily Roger was in gracious mood and merely patted me benignly on what we will call, for reference purposes, my hair.

But what of the man Thorogood and his band? I hear you lisp. Well, he had your Uncle John coming as close to dancing as he ever gets, and the perspiration was fair cascading down my body by the end of the set. Good, rowdy, vulgar stuff. It is a fact, you know, that those old blooze riffs never die – although Blast Furnace, who was also making an exhibition of himself at the front of the stage, does what he can to stifle the buggers – and such diverse persons as Bootsy Collins, who shook me warmly by the hand and said 'Good to meet ya. Jes' checkin' it out' (does this count as an interview, sweet Ed?) and Nick Lowe (you know!) were alongside me on the floor.

Now, let's talk some more about football. (*No chance, fatty! Ed.*)

# World Service

*Guardian*, 17 July 1996

I HAVE a friend who recently had a heart transplant. If you met him now, so healthy does he look, and indeed is, that you would think you need a heart transplant too. It's quite clearly nonsense; and that's what I think when I read about the plans to reform the World Service.

Just because these policies work in other areas of the BBC, it doesn't necessarily mean they will work in the World Service. We risk ending up with 'lean machine' radio that is a bit like the bionic man: rather heartless. With the World Service, you are dealing with intangibles like soul – things you can't put down into your accounts and can't rationalise out in policy documents. A sentiment exists, something that is anathema in contemporary business culture. I never feel entirely happy about centralisation, however it's presented; and what I have read about these reforms suggests to me that we are setting upon a potentially dangerous path.

I speak as a bloke who has been on the World Service double-decker bus which used to trek around eastern Europe bringing the good news about the service to recently liberated countries. We'd pull into little market towns in Bulgaria and considerable crowds would turn out to greet us. Hundreds of photos of me were handed out as a symbol of what the BBC was – and whereas here they'd end up on the street, there for some reason they were anxious to embrace any they could get their hands on.

It would be depressing to rationalise away that degree of commitment from your audience. My own programme gets a tremendous response from its listeners, and it is incredibly moving to receive postcards from people half-way up a mountain in a remote part of the world. All I'm doing is presenting a programme of noisy records – and if it has such an effect and makes people feel at home, how much more profound must be the effect of the more serious programmes?

# Robert Wyatt

*Sounds*, 3 August 1974

---

IF THE bulk of your education comes, as does the bulk of mine, from a five-minute romp through the *Daily Mirror* every morning, you're probably of the opinion, as I am, that self-denunciation is an important part of daily life behind the Iron Curtain.

Stand up and tell an astonished world about your moral degeneracy and how you have unwholesome lusts after the lathe-operator from Kollectiv G. That's the ticket – and you'd imagine that an exciting Radio 1 DJ would seldom find himself in a situation where self-denunciation was essential to forestall denunciation by a pack of marauding musicians.

Come with me, if you will, in your fertile imaginations, to the party held at no great distance from Eel Pie Island, to celebrate the wedding of Robert Wyatt to Alfreda Benge. I had gone there with a pile of African records to lend Robert and the promise of an impending fondue set with which to dazzle the newly-weds.

Their flat was crammed with relatives and practically everyone who ever has or ever will record for Virgin Records. Here was label chief Richard Branson, here was promo man Al Clark, over there A&R wallah Simon Draper.

Having paid my compliments to our host and hostess, I went with yummy Radio 1 producer and trick cyclist John Walters, the

Pig and Pig's smallish sister, to sit on the grass and tell sad stories of the death of kings, you know, how some were sleeping killed and so on.

No sooner, no sooner, I tell you, than I'd lowered my massive frame on to the sward when I overhead someone saying in a loud voice, 'John Peel was a bit odd.' Now, being called 'a bit odd' when you're a Radio 1 DJ can be regarded as tantamount to high praise, so I was just stoking up for a modest blush when a clamour of embarrassed voices warned the speaker, before he pressed on with further insights into my character, that I was lounging just behind him. He turned and came over to sit beside me.

It transpired that he was – and still is, I shouldn't wonder – Anthony Moore of Slapp Happy. Now Slapp Happy were on a recent *Top Gear* – they will be on again – and were assisted on their session by the very same Robert Wyatt whose nuptials we had gathered together to celebrate. Anthony felt – and in retrospect I think he's right – that I had been rather non-committal about their music. Flippant even, uncaring.

Fred Frith of Henry Cow, who was sprawling near by, added his voice to the clamour, saying that he felt that I'd been rather cavalier in my approach to their last session. Reeling under this twin assault, I was floored when the man Walters, from behind a huge mound of food, opined that I had been clumsy with recent numbers from Na Fili and Swan Arcade. They had – and I remember them well – recorded a couple of very thoughtful, moving numbers for the programme and John, 'Petals' to his friends, told me that he hoped neither band had heard the insensitive way I had back-announced their work.

Thinking on't I had to agree with him and this led to an agony of self-denunciation in front of my critics. The trouble is, roughly, that for a long time John Peel was regarded as something of a pundit. There was a time – and this'll make you laugh – when folks advertising for pen-friends would describe themselves as

'Peelites'. Now that's a difficult position for a simple country lad with four 'O' levels to find himself in.

However, I tried to do programmes that justified this faith, to make comments on the music that were thoughtful if not profound, honest if not well-considered. As a result, in my view, the programmes became rather like *Stars On Sunday*, skating over the surface while creating an illusion of depth. Also other programmes in similar vein threatened to be even more boring than my own. So I elected to try to move from this uncomfortable, actually unworkable, posture to one in which I could play the same music while attempting to convey the impression that good music need not necessarily be presented in a morbid context. Obviously I've gone too far, have become frivolous, so the pendulum will have to swing back slightly.

In the meantime, I apologise to Slapp Happy, Henry Cow, Na Fili, Swan Arcade and others I may have unwittingly offended. I'll endeavour to mend my ways and to deal with more sensitive music in a more fitting way. Trouble is, I only find out where I'm going wrong when someone else points it out. How much easier it would be to do a straight Top 40 programme, loads of laffs and jingles.

Ah well. See you next week, boys and girls. Don't forget to keep your nails clean.

# Lena Zavaroni

## More Than Enough Music to Go Around

*Sounds*, 20 July 1974

---

FORSOOTH AND odds bodikins, but we music freaks are an ill-assorted and motley crew. Lewd fellows of a baser sort, some aver, but I don't go along with that. We are cantankerous though, cantankerous and short-sighted, frothing mightily with absurd prejudices and liable to lash out at the lowering of a chapeau if anyone saunters across any one of those prejudices.

So my initial reaction to a letter from a Gareth of Bristol was rage and muffled cursing. Gareth railed, in a surprisingly retiring way, against my snide remarks, on radio, about Lena Zavaroni and members of the Osmond family. 'Let 'em be,' was the tenor of his remarks, 'for they bring pleasure into young lives.'

Then in last week's ZOUNDS there was a letter or two from disgruntled readers hinting that my enthusiasm for the current Slade single indicated that I was suffering from senile decay, hardening of the arteries, halitosis and spots. How, they wanted to know, could I enjoy Slade but not enjoy ELP, Yes and Focus and still call myself a member, albeit in poor standing, of the human race?

So last night, as I steered my motor-car through the streets of London with all the delicacy of touch of a Johann Cruyff, I listened again to an 8-track of *Topographic Oceans* in the hope that either (1) I would understand why Yes are so highly regarded by the readers of this paper or (2) I could finally put my finger on just what it is about the band that I dislike.

Unfortunately neither revelation came my way. I was left, as before, unmoved – although mightily impressed with the skill that the Yesses bring to their several instruments.

However, ruminating as I cruised in search of a place in which to rest the night, I thought about bands, tastes, readers' letters and prejudice and came to the conclusion that my constant bitching about Zavaronis and the ELPs of this world is due not so much to out-and-out blind loathing of what they do, are, stand for; but rather frustration that so much attention is focused on them to the exclusion of bands, musicians, performers who are, in my view, worthy of a portion of that attention.

Yes, that's it.

The programmes I do for Radio 1 have always been (roughly) based on the principle that what you're buying, listening to and enjoying is all very well but there exists also something else, less favoured, but equally worthy of your attention. I mean, there's a guitarist in a Peruvian band (and I'm not making this up) who's a knock-out – and who knows what marvellous bands there may not be in, say, Poland or Zaire or Iceland. I want to know about them – and to let you know about them.

Last night's *Top Gear* was, for me, one of those programmes that inexplicably takes off in such a way that getting back down again when it's all over is really a difficult, even unpleasant, task. We started with a track from a Capricorn LP by Grinderswitch (a fine new band) which featured Dickie Betts and moved on through Bryn Haworth, Sandy Denny, Magma, Don Covay, Cream, Sweet (yes, Sweet), Ry Cooder, The Fatback Band, Eric

Clapton, Willie Mitchell, Bill Black's Combo, Kevin Coyne, Joni Mitchell and Sunny Ade and His African Beats (from Nigeria).

That covers, I think you'll agree, a pretty wide range of musical tastes – no Zavaroni, no Focus, Yes or ELP, I'll admit, but I suppose I should just own up and admit that I can't come to terms with their music and leave it at that. There's not enough time for all this bitching and the bitching can obscure the fact that, almost uniquely among our remaining resources, there's still plenty of good music to suit all tastes.

Actually, we had a competition on *Top Gear* to identify the Sweet track – the 'B' side of their current chart single – and we got about 300/400 replies. About fifty listeners got it right, the remainder split their guesses between, among others, Status Quo, Slade, Jeff Beck, Led Zeppelin, Mud, Rod Stewart, The Heavy Metal Kids, The Sadistic Mika Band, Deep Purple, Queen, Stray, Uriah Heep and the Pink Fairies.

Possibly some of the guessers will have been mildly embarrassed to find that the record (which is excellent, by the way) is by Sweet. I suppose the contest was evolved with that in mind. On the other hand I suspect most listeners are grateful for good music from whatever source and from the lists of requests that are sent in to *Top Gear* it's clear that many people have much broader tastes than the BBC imagines.

I got one the other day that asked for a record by either Ron Geesin or Charlie Feathers – and you don't get much further apart than Ron Geesin and Charlie Feathers.

Perhaps then it's time for me to stop writing a load of bad-tempered badger-piss about other people's tastes and for some ZOUNDS correspondents to do the same. As I said, there's more than enough music to go round.

# Picture Credits

1: © Brian Moody/Rex Features; 3 bottom right: © Ray Stephenson/Rex Features; 4 bottom: Jill Furmanovsky; 6: © Claudine Schafer

# Index

A&M 209
AC/DC 30–1
*Adam and Joe Show, The* 338
Adorable 101
Aerosmith 245, 248
African music 15–16, 26, 177
Airplane 14
Alaap 22–3
Albert Hall 214–17, 278
Albini, Steve 290
*Alien Empire: Replicators* (TV programme) 4–5
All American Solid Silver 60s UK Tour 62
American rock 183–4
Amin, Idi 106
Amon Duul II 138
*Amongst the Catacombs of Nephren-Ka* (album) 18
Anka, Paul 58
Apollo Theatre (Manchester) 237
*Archers, The* 7–8
Arm, Mark 291
Armatrading, Joan 13
Arthur, King 131–2
Ascended Masters 53
Ash Ra Tempel 136, 137–8
asteroids 53
Auteurs 101
Autry, Gene 235
Ayers, Kevin 312
Ayles, Martin 308

Bad Boys Inc 240
Badenough, Boris 42
Baez, Joan 178, 190
Baker, Clarence 160
Baker, Danny 242
Ball, Zöe 337
banned bands 180–5
Bannockburn, Timmy (Tony Blackburn) 28, 166–7
Barton, Geoff 228
Bates, Simon 240, 314
Bay City Rollers 28, 176, 177
BBC Radio Lancashire 144
BBC Radio Nottingham 144
BBC World Service 58, 158, 345–6
Be-Bop Deluxe 14, 177, 179
Beardsley, Peter 131
Beatles 14, 57, 152, 184, 267
Bedford, David 312
Beefheart, Captain 25, 36–8, 40–1
Bembeya Jazz National 15–16

Benge, Alfreda 347
Benji, Johnny 76
Bennett, Tony 17–18
Berlin punks 19–21
Berry, Chuck 140
Betts, Jo 86
bhangra 22–4
Bhundu Boys 15, 25–6, 154
Big Black 291
*BIKE* magazine 304–10
Bjork 243, 244
Black Crowes 225
black music 176
*Black Seeds of Vengeance* (album) 18
Black Wax 212–13
Blackburn, Tony (Timmy Bannockburn) 112, 127, 163
Blackwell, Nigel 108, 109
Blast Furnace 343
Blue 249
Blueboy 281
Bluetones 283
Bobcat, DJ 238
'Bohemian Rhapsody' 29
Bolan, Marc 115
Boney M 58
Bonzo Dog Band 109, 284, 285, 289
Boo Radleys 242
*Book of the Archers, The* 7
Boorstin, Daniel J. 175
Bore category 28–9
Bowie, David 114–15, 178, 267
Boy George 239
*Boy Meets Girl* (radio programme) 288
Brady, Paul 102
Bragg, Billy 102, 239
Brand X 222
Branson, Richard 347
Bread 37, 95, 184
Breedwell, Brooke 6
Brinsley Schwarz 178
Bristol University 86–7
British Forces Broadcasting Service 19
Bronner Brothers 213
Bronski Beat 239
Brookes, Bruno 158, 159
Bros 159
Brotherhood of Man 70, 250
Brown, Arthur 267–8
Bruce, Ken 279
Bruno, Frank 123
Budd, Kate 66
Burning Spear 212
Burton, James 61
Burton, Richard 29
Butthole Surfers 32–3, 291
Bygraves, Max 267

California 35
Callister, Stephanie 314
Campbell, Cornell 212
Can 138
Capital Radio 343
*Captain Fantastic and the Brown Dirt Cowboy* (LP) 178
Carducci, Antonio 53–4
Carling, Will 45
Carpenters 72, 267
*Cashbox* magazine 71
Cassidy, David 218, 219, 267
Chapin, Harry 176
Charles, Prince 92–3
chat shows 163–4
Château de Castel Novel (Varetz) 55
Cheese, DJ 316–17
Chelmsford punk festival 182
children's television 44–5
Chilli Willi and the Red Hot Peppers 178
China Drum 243
Chuck D 238

Cirith Gorgor 187
Clapton, Eric 62
Clark, Al 347
Clarke, John Cooper 184
Clash 184
Claw Boys Claw 90
Clemence, Ray 104
Cluster 137
CNN 225
Cobain, Kurt 223
Coda Records 171
Cohen, Leonard 41
Collingwood, Charles 7
Collins, Boosty 343
Collins, Phil 221
Columbia Records 184
comedians 287–8
Como, Perry 267
Connolly, Billy 287, 288
Conteh, John 178
Cooder, Ry 61, 195
Cope, Julian 224
Cotton, Bill 288
Cougars 206
country music 177
Cow, Henry 312
Coxhill, Lol 312
Crane's (Liverpool) 252
Cream 14, 59
Creations, The 213
Crickets, The 62
Cropper, Steve 61
Crosby, Bing 184
Culture Club 239
Currie, Alannah 302
Cut Creator 238

*Daily Express* 177
Dalglish, Kenny 47–8, 85, 227
Dalziel, Lt Col Sylvia 50
Damned 182
Dann, Trevor 172
Davidson, Jim 5
Davies, Gary 123, 143, 240
Davies, Sir Peter Maxwell 278
Davis, Paul 111
Day, Mark 111
death metal 18
Deayton, Angus 46
Deluxe, Nitro 42
Denver, John 72, 176
Devera Ngwena Jazz Band 15
Devo 184
Dexter, Jeff 51
Diabate, Sekou 16
Diamond, Neil 176
'Diddy Wah Diddy' 36
Die Cheerleader 225
Dirty Rotten Imbeciles 170
*Disc* 39–41, 88–9, 94–6, 119–21, 141–2, 146–50, 197–200, 214–17, 260–2, 273–7, 327–30
disco scene 176–7
Disley, Terry 172
Disposable Heroes of Hiphoprisy 224
Dodd, Jeggsy 108
Dodgy 102
'dog-sniff' 20
Donaldson, John 269
Donovan 103, 198
Doobie Brothers 250
Doomsday 52–4
Dr Jeckyll and Mr Hyde 317
Dransfield, Barry 284
Draper, Simon 347
Dreadzone 283
drugs 295–6
Duke Duke and the Dukes 179
Dunlop, Joey 307
Dunlop, Robert 306
Duran Duran 239
Dylan, Bob 71, 178, 179, 196

Eagles 176
East 17: 240

Eastern Europe 57–60
Eat 102
Eddy, Duane 61–3, 195
Ekland, Britt 7
Elastica 282
Electric Chairs 249
elfshot 156
Elizabeth, Queen 198
Elliott, Jack 235
Ellison, Tommy 213
ELO 89
ELP (Emerson, Lake and Palmer) 181, 205, 206, 220–1, 222, 225, 311
Emerson, Keith 221
Empty 291
England vs. Argentina match (1977) 104
Enid, The 249
Erskine, Peter 88–9
Escalades (Spain) 56
Escargot-Peyle, Simon d' 141
Escher, M. C. 137
Esquires Ltd 213
Europe 64–5
European Cup Final (1978) 227
Eurosonic 187
Eurovision Song Contest 39, 66–8, 69–72, 168
Extreme Noise Terror 75–6

Faces 14, 25, 88, 89, 205
Fairport Convention 13
Faith No More 225
Fall 91, 241, 269, 315, 338
Farley Jackmaster Funk 42
Fastway 30
Fatima Mansions 223
Faust 138
Feathers, Charlie 352
Ferry, Bryan 92–3
festivals 294 *see also* individual festivals
'Final Countdown, The' 64, 65
Fisk, Steve 290
Five Hand Reel 246–7
Flavor Flav 238
Fleetwood Mac 183
flies 4
Flinch 243
floods 53
Focus 205
Fogarty, Carl 309, 309–10
Foo Fighters 244
football 94–5, 270
Forgan, Liz 278
Foster-Moore, John 10
fox hunting 97–9
Frankie Goes to Hollywood 240
Franklin, Aretha 49
Franklin, Revd Mr 49–50
Freeman, Alan 13, 221
Friche, Florian 138
Frith, Fred 348
Frith, Simon 9

Gambaccini, Paul 209
Garfunkel, Art 176
Garner, Ken 284
Gates, David 37
Gaughan, Dick 246, 247, 250
Gee, Jakki 314–15
Geesin, Ron 352
Genesis 71, 92, 181, 221
Gibberish G 75
Gibson, Jim 268–9
Glastonbury festival 17, 100–3, 294
Glitter, Gary 218, 219, 267, 333
global warming 130
God Machine 223
'God Save the Queen' 106–7, 180, 181, 321
Golden Earring 246, 247
Golden Flask 166–7
Gong 312

Goo, Hiromi 71
*Goodies, The* 287
Grandmaster Flash 317
Grateful Dead 13, 14, 219
Greene, Patricia 7
Grinderswitch 351
Grohl, Dave 244
Groundhogs 178
Grundy, Bill 21
*Guardian* 7–8, 47–8, 100–3, 107, 223–5, 239–41, 242–4, 281–3, 284–5, 345–6
Guthrie, Woody 235

Half Man Half Biscuit 108–9
Hammersmith Palais 9
Hands 247
Happy Mondays 110–11
Hardcastle, Paul 301–2
Hardin, Eddie 171
Harley, Steve 29
Harper, Roy 179
Harris, Bob 73, 275
Harris, Mick 76
Harris, Rolf 100
Harrison, George 61, 151, 152–3, 195, 196
Hartman, Dan 302
Harvest label 179
Harvey, Alex 250
Haugland, Ian 65
Hawkes, Mike 283
Hawkwind 250
Haynes, Gibby 33
Hayward, Nick 333
Hazlett, Steve 308
Heera 23
Hefner 187–8
Hellawell, Keith 296
Hendrix, Jimi 51, 110, 230–1
Henri, Adrian 115–16
Henry Cow 348, 349
Her Liberty Orchestra 207
Heresy 169
Hill, Jimmy 270
hip hop 237, 316–17
hippies 117–18
Hislop, Ian 2, 46
Hislop, Steve 308–9
*Hissing of Summer Lawns, The* (LP) 179, 210
Hobbits' Garden (Wimbledon) 92
Hoffman, Dave 143
Hole 223–4, 242
Holland 186–7
Holle Holle 22, 23
Holmes, Eamonn 189
Holy Terror 170
Home 95
Hoodoo Rhythm Devils 198
House music 42–3
Howe, Alison 51, 283
Howlett, Kevin 190
Hughes, Emlyn 320
Humble Pie 178
Hurley, Steve Silk 42
Hurll, Michael 301

Incredible String Band 14
*Independent on Sunday* 57–60, 61–3, 268–9
Intone Records 212
Ipswich 75, 119–21
Island Records 179
Isle of Man 130, 304–10
Isle of Wight Festival 51, 293–4

Jack the Lad 263
'Jack Your Body' 42
Jackson 5: 198
Jackson, Bob 308
Jackson, Chrissie 315
Jackson, Michael 122–4, 198
Jacob's Mouse 223
James, Clive 189

'Jealous Guy' 93
Jefferson Airplane 51
Jefferson Starship 176
Jennings, Waylon 177
Jensen, Kid 79
Jesus and Mary Chain, The 9, 10
Jethro Tull 176
Jewell, Dick 343
JFM 144
J. M. Lulu 213
Joel, Billy 58, 125–6
John, Elton 72, 176, 178, 184, 334
*John Peel's Music* 19
Johnson, Brian (Beano) 31
Johnson, Holly 239–40
Jones, Dean 76
Jones, Linda 207
Jones, Steve 191
Joystrings 50, 51

K 291
Kamath, Anita 18
Kane, Ray 142
Kate Brothers 209–10
Keegan, Kevin 47
Kerguelen 127–9
Kershaw, Andy 26, 101, 102, 239, 240, 304, 309
Khan, Morgan 316
Kingdom Come 206
Kingfish 246
Kirkpatrick, John 284
Kissinger, Henry 29
KL FM 143
Klee, Paul 328
KMEN 36
Knebworth 135
*Knocking at Doomsday's Door* (TV programme) 53
Knuckles, Frankie 42
*Kojak* 29
Komintern 206
Komische Musik 136–8
Kozdra, Wladyslaw 57
Kraftwerk 42

Ladysmith Black Mambazo 213
Lama, Serge 71
Lamacq, Steve 243
Lamarr, Mark 340
Laine, Frankie 139–40
Last, James 176
Lauper, Cyndi 143
leaf-hopper 5
Leary, Timothy 137
Led Zeppelin 72, 181
Lennon, John 176, 184
Levellers 5 269
Lincs FM 144
Lindisfarne 178
Lindley, David 61
*Listener, The* 69–72, 136–8, 175–9, 180–5, 218–19, 266–7, 286–9, 311–13
Little Feat 25, 179
Little River Band 248
*Live at the Counter-Eurovision (79)* (LP) 160
*Live and Kicking* 46
Liverpool Empire 139
Liverpool FC 39, 47–8, 131, 142, 193, 227, 320
Liverpool Scene 116
Liverpool University 114–15
Living Colour 225
*Living with the Enemy* (TV programme) 258
LL Cool J 238
local radio 143–5
Logan, Johnny 67, 68
Loggins and Messina 176
Lone Star 246
Long, Janice 302, 303
Love Birds 213
'Love Can't Turn Around' 42
Love, Courtney 223, 242

Lovelace Watkins 146, 148–50
Low, Andy Fairweather 209
Lowe, Nick 344
Lowry, Shauna 44–5
Lupu, Radu 216
Lynne, Jeff 195, 196
Lynott, Phil 250

McCallen, Phillip 308
McCarn, David 236
McCartney, Paul 61, 195
McDiarmid, Mac 307
MacDonald, Dr David 98
McGough, Roger 115
Madonna 151–3, 154–5, 239
Mahotella Queens 213
Maldita Vecindad 100–1
Mamonov, Peter 319
Manjeet 23
Manor House Ballroom (Ipswich) 119–20
Marcello, Kee 64
Margaret, Princess 7–8
Marley, Bob 177
Marr, Johnny 276, 277
Marsh, Sam 223
May, James 2
Mayall, John 59
mbalax 16
Medicine Head 263
medieval medicine 156–7
*Melody Maker* 178, 181
Members, The 213
Menswear 281
Mercury Rev 225
Mettingham Castle 118
Michael, George 239, 333–5
Michaeli, Mic 65
*Mighty Gorgon* (LP) 212
Miles, John 248
Miller, Frankie 249, 250
Minier, Christine 68
Minogue, Kylie 158–9
Misty in Roots 160–2
Mitchell, Joni 176, 178, 179
Money, Zoot 284
Montel 164–5
'Moon Child' 37
Moore, Anthony 348
Moore, Ray 68
Morecambe and Wise 287
Morgan, Nick 86
Morris, Mixmaster 103
Morrison, Van 36, 102
Morrissey 276–7
Morton, Tom 50
Motors 182, 249
Mountain Sisters 213
Moyo, Jonah 15
Mud 267
Mudhoney 291
Mungo Jerry 243
Mustaphas 161
MX-80 Sound 184
My Bloody Valentines 58
My Generation 293, 294
*My Top Ten* 47

Na Fili 348, 349
Napalm Death 169–70
Nation, Trevor 306, 308
*Neighbours* 159
Neu 138
New Age music 171–2
*New Musical Express* 181
New Year's Eve 173–4
New York Dolls 178
Newport Folk Festival 190
Nightingale, Anne 240
Niklaus, Hedli 7
Nile 18
Nirvana 223
Noblett, Michael 310
Noiseville 268–9
Noorderslag 187–8
Nugent, Ned 31

*Observer* 9–11, 15–16, 22–4, 25–6, 30–1, 32–3, 36–8, 42–3, 64–5, 66–8, 75–6, 90–1, 92–3, 108–9, 110–11, 122–4, 125–6, 139–40, 151–3, 154–5, 158–9, 160–2, 169–70, 171–2, 195–6, 233–4, 235–6, 237–8, 255–6, 276–7, 290–1, 301–3, 314–15, 316–17, 318–19, 333–5
*Oddballs* 189–90
O'Donoghue, Daniel 86
Oladunni Decency 207
*Old Grey Whistle Test, The* 182–3, 343
Oldfield, Mike 176, 178, 312–13
'Only the Light' 66–7
*Ooh La La* 88–9
Orbison, Roy 140, 195–6
Original Blind Boys 207
Osmond, Donny 219
Osmonds 198, 199, 201–2, 203, 205, 207, 218, 219, 267
Otis 46
Overdrive, Bachman-Turner 176

Paisley, Robert 320
Papa and the Utopians 213
Parker, Graham 228, 248
Parkinson, Michael 29
Parris, Matthew 127–9
Patten, Brian 115
Pavitt, Bruce 291
Peebles, Andy 144
Peebles, Rikki 66–7, 68
Peel Roadshows 80–2, 260–2, 263–5
Peel's Theory 219
Penn, Sean 151, 152
People Unite Musicians' Cooperative 160
Père Ubu 184
Perfect Daze 76
Periodical Publishers Association awards (1998) 1–2
petrol rationing 202–3
Pettifer, Julian 98–9
Petty, Tom 196
Phoenix Festival 223–5
photo competition (*Sounds*) 77, 82–3, 86
Pink Floyd 25, 71, 72, 89, 137, 176, 178, 179, 205, 206
Pink Pop Festival 228–9
police 192–3
Police (band) 243
Pop Will Eat Itself 225
Popllama 291
Popul Vuh 136, 138
Powell, Enoch 207
Powell, Peter 239, 240
Premi 23
Presley, Elvis 184, 266
Prince 233–4
Prince, Tony 209
Prior-Palmer, Lucinda 7
protest music 235–6
Public Enemy 237–8
Pulp 283
*Punch* 112–13, 251–4
Punjabi Pop festival 22–4
punk/punk rock 19–21, 180–1, 182, 183, 184
Pursey, Jimmy 184

Q103 FM 143–4
Quartz 249
Quatro, Suzi 267
Queen 29, 176, 177

Racing Cars 249
Radio 1: 13, 47, 183, 239
Radio 1 Stanley 94
Radio Borders 145
Radio London 115

Radio Mafia 281
*Radio Times* 1–6, 17–18, 19–21, 34–5, 44–5, 49–51, 52–4, 55–6, 73–4, 97–9, 114–16, 127–9, 130–2, 143–5, 156–7, 163–5, 173–4, 186–8, 189–90, 257–9, 278–80, 292–4, 295–6, 299–300, 324–6, 331–2, 336–8, 339–40
Rantzen, Esther 163
Ravenscroft, Alan (brother) 98, 278–9
Ravenscroft, Alexandra (daughter) 8, 44, 85, 174
Ravenscroft, Flossie (daughter) 44, 174
Ravenscroft, Sheila (wife) 2, 17, 34, 74, 84–5, 118, 214, 299
Ravenscroft, Thomas (son) 44, 173
Ravenscroft, William (son) 12–13, 44, 100, 104, 114, 173
Read, Mike 240
Reading Festival 30, 182, 194, 242–4, 245–50
Real Sound 15
record shops 251–4
records' parties 208–9
Reed, Jimmy 209
reggae 177
Reggie's Soweto Magic Band 213
Reynolds, Stanley 107
Rhys-Jones, Griff 325
Rice, Anneka 7
Richard, Cliff 255–6
Richard, Keith 277
Ricotti, Frank 312
Ridgley, Andrew 334
Ritchie, Lionel 47
road rage 257–8
Roberts, Andy 115, 116, 284
Roberts, Trish 315
Rock Against Racism 161
Rock FM 144–5
rock music 137, 175, 266–7, 311
rock stars 175–6, 178
*Rock That Doesn't Roll, The* (TV programme) 49–50
Rockingbirds 102
Rockpool 268–9
Rods 246, 247
Roe, Tommy 62
*Rolling Stone* 177
Rolling Stones 13, 58, 72, 176, 184, 218–19, 267
Ronson, Mick 178
Ronstadt, Linda 183–4
Roslin, Gaby 336, 337, 338
Ross, Diana 140
Rotten, Johnny 180
Roxy Music 13, 89, 92, 176
Royal Albert Hall see Albert Hall
Royal Holloway College 204–6
Royal Liverpool Philharmonic Orchestra 216
Rumour, The 228, 248
*Rumours* (LP) 183
Rushton, Willie 324
Russell, Ken 297
Russell, Rosalind 89
Russian groups 318–19
Ryback, Timothy *Rock Around the Bloc* 57, 59
Ryder, Paul 111
Ryder, Shaun 111

Sadistic Mika Band 179
*Safe as Milk* (album) 37
Salt 246
Salvation Army 50
Saunders, Jessie 42
Savalas, Telly 29
'Save Your Kisses for Me' 71
*schiffsbergrüssungsanlage* 19, 21
Schulze, Klaus 136, 137
*Scorn* 127

Scott, Jake 202–3
Scott, Roger 343
Scratch Acid 290
Seattle 290
Secret Shine 281
Self, Laurie 174
Serrat, Joan Manuel 71
Sex Pistols 21, 106–7, 180, 181, 184, 288, 321
SGR FM 143
*Shabini* (LP) 26
Sham '69 184
*Shanghai Surprise* 151
Shannon, Del 62
Shannon, Sharon 102
Sherwood, Adrian 23
Shrapnel, John 4, 5
Shrewsbury 112–13
Simon, Paul 176
*Situation Vacant* (TV programme) 324, 326
sixties 292–3
Slade 350
Slapp Happy 348, 349
Slits 184
Smith, Harvey 178
Smith, Mark E. 91, 269, 338
Smith, Patti 28, 177–8, 184
Smith, Tommy 105, 227, 320
Smiths, The 276–7
Snafu 263, 264
Softs 14
Sokolov, Grigory 280
Somerville, Jimmy 239
Son House 328
Sonar Festival (Barcelona) 55–6
*Songs in the Key of Life* (LP) 183
Sonic Youth 9–10, 224, 290, 291
Soul Twins 213
Soundgarden 291
*Sounds* 12–14, 27–9, 32, 77–9, 80–3, 84–7, 104–7, 117–18, 133–5, 166–8, 191–4, 201–3, 204–7, 208–10, 211–13, 220–2, 226–9, 230–2, 245–50, 263–5, 270–2, 297–8, 320–3, 341–4, 347–9, 350–2
Southall riots (1979) 160
Soviet Union 318–19
Spandau Ballet 239
*Spitting Image* 131
Spottiswood, Cherry 117
Springsteen, Bruce 28, 177
Staa Marx 246
Stafford, Jo 139
Stanshall, Viv 109, 284–5, 289, 312
*Star*/Babycham Female DJ of the Year Competition (185) 314–15
Starr, Ringo 176
Status Quo 306
*Steal You Face* (LP) 13
Steeleye Span 176
Stevens, B. J. 315
Stevens, Jimmy 142
Stewart, Ed 287
Stewart, Rod 176, 178
Stock, Aitkin, Waterman 159
Stowmarket 74
Stranglers 182
Street Sounds Records 316
Streisand, Barbra 183
Sturgess, Claire 240, 282
Sub Pop 290–1
Suede 101
Sumlin, Hubert 74
*Sunday Mirror* 106, 107
*Sunday Times* 286–7
Superchunk 101
Supergrass 281
Supertramp 176
Swan Arcade 348, 349
Swarbrick, Dave 284
Sweet 352

T. Rex 115
'Talk of the Town, The' 146–8, 198
Talking Heads 184, 277
Tangerine Dream 136, 137, 179, 206
Tarrant, Chris 324
Taylor, Elizabeth 29
Television 184
Tembo, Biggie 25–6
Tempest, Joey 64, 65
Terminator X 238
That Petrol Emotion 224
Theakston, Jamie 337
Them 36
Themis, John 172
Thin Lizzy 177, 245, 248
Thomas, Nicky 160
Thompson, Danny 284
Thompson Twins 302
Thorogood, George 343
*To the Ends of the Earth: Dreaming on Desolation Island* 127
*Today* (radio programme) 295–6
*Tommy* (film) 297–8
*Tomorrow's Today* (LP) 171
Tonderai, Mark 240
Tong, Pete 240
*Top Gear* 351–2
*Top of the Pops* 9, 29, 181, 301–3
*Top of The Pops* (magazine) 2
Townshend, Pete 62
Toye, Lori Adaile 53
Traffic 178
Traveling Wilburys 196
Travis, Dave Lee 158–9
Tremeloes 58
*Trick of the Tail* (album) 221
TT Races 304–10
*Tubular Bells* (album) 312–13
turntablism 34
*Twang's the Thang, The* (LP) 61
*Twenty-first Century Fox* 98–9

U-Boat 246
UK Fresh 86: 316–17
Ultravox 247–8
*Under the Sun: Painted Babies* 6
Undertones 91
Uriah Heep 246, 247

Van Vliet, Don 36, 37–8, 40–1
Vane, Phil 76
Vee, Bobby 267
Velvet Underground 101
Venue for Ipswich Campaign 75
Verve 102
Viking FM 144
Vincent, Gene 206, 267
Vinton, Bobby 267
VIP 44–6
Virgin Records 179
voice-overs 324–6

Wainwright, Loudon 13, 327–8
Waits, Tom 28
Wakeman, Rick 178, 221–2
Walker, Scott 95
Walsall 80–1
Walters, John 77, 183, 275, 284, 289, 331–2, 339, 342, 347–8
Ward, Michael 267
Waterman, Pete 159
Waterpop free festival (Wateringen) 90–1
Weddoes 225
Weir, Bob 275
West Midlands College of Education 80
Westminster, Duke of 7
whaling 190
Wham! 334–5
*Whatever You Want* (TV programme) 336–8
Whelan, Gary 111
*Where It's At* (TV programme) 70
White Stripes 340

Who 13, 72, 205
Whole World 312
Wilde, Kim 123
Wildings (Shrewsbury) 251
Williams, Fay 315
Williams, Ray 213
Wimbledon 132
wine-tasting 342–3
Winwood, Steve 312
*Wish You Were Here* (LP) 178, 179
Wogan, Terry 7
Womack, Bobby 277
*Women Who Kill* (TV programme) 259
Women's Institute 339–40
Wonder, Stevie 183
Word of Mouth 316
World Cup 270
Wray, Link 62
Wright, Steve 240
Writing On The Wall 202–3
Wyatt, Robert 275, 312, 347–8
Wynette, Tammy 177

Yardbirds 14
Yates, Paula 314
Yes 181, 205, 225, 311, 351
Young, Angus 30–1
Young, Malcolm 30
Young, Neil 179

Zavaroni, Lena 350
Zdob Si Zdub 187
*Zeit* (LP) 137, 206
*Zuma* (LP) 179
Zvuki Mu 59, 319